T0277315

THE ART OF
WAR IN
TWENTY BATTLES

THE ART OF WAR IN TWENTY BATTLES

ANTHONY TUCKER-JONES

For Amelia, Henrietta and Ophelia
who make it all worthwhile

First published 2018 as *The Killing Game*
This paperback edition first published 2023

The History Press
97 St George's Place, Cheltenham,
Gloucestershire, GL50 3QB
www.thehistorypress.co.uk

© Anthony Tucker-Jones, 2018, 2023

The right of Anthony Tucker-Jones to be identified as the Author
of this work has been asserted in accordance with the
Copyright, Designs and Patents Act 1988.

All rights reserved. No part of this book may be reprinted
or reproduced or utilised in any form or by any electronic,
mechanical or other means, now known or hereafter invented,
including photocopying and recording, or in any information
storage or retrieval system, without the permission in writing
from the Publishers.

British Library Cataloguing in Publication Data.
A catalogue record for this book is available from the British Library.

ISBN 978 1 80399 380 5

Typesetting and origination by The History Press
Printed and bound in Great Britain by TJ Books Limited

MIX
Paper from
responsible sources
FSC
www.fsc.org   FSC® C013056

Trees for Life

CONTENTS

CACOPHONY OF WAR

There is always a racket, often drums, chanting or just plain banging. The trick is, you get your game face on and psych out your opponent and that often involves noise – a lot of noise. It is all a game really. You have to wage psychological warfare against your enemy before the fighting even starts. You have to convince your enemy that you are more likely to kill him, than him kill you.

Just before battle a soldier will get a nagging knot in the pit of his stomach, even feel sick at the thought of dying or killing another human being. The taste of bile is never pleasant. There is no consolation save gripping your weapon even tighter as if it were some lucky talisman. When the adrenalin cuts in you have to channel it into fight, not flight. Running is for cowards.

It is the opening noise of battle that first tests a man's courage. While waiting for the enemy there will be the steady tramp of thousands of feet, or worse still the pounding of thousands of feet charging towards you. Loud rumbling might herald the approach of chariots full of spearmen and archers; or thunder signal galloping hooves and cavalrymen armed with lances.

Once an army is deployed, the soldier has to endure the whistle of arrows or musket balls, or the whizz of cannonballs flying through the ranks arrayed around him. The soldier's lot in life is to endure all this, for he cannot move without orders, if he does then desertion is always punishable by death. All this carnage and mayhem before the soldier has even got into battle.

He will see the distant flashes followed by the boom of the guns. 'Pray don't let it be me,' thinks the soldier, 'please don't let it be me – or if it is, make it quick. Don't let something be torn off followed by searing agony.' No soldier wants to die listening to his own screams as his blood ebbs away in the dirt. That is a cruel end, likely to make his brothers-in-arms throw down their weapons and flee in blind terror. In some instances, though, death is preferable if it means not being dragged off to the temple and having your still-beating heart torn from your chest. Never listen to the cacophony of war, only listen for orders and press on, no matter what happens.

In 1879 British soldiers could hear a terrible rumble, like an oncoming freight train, it heralded the onslaught of 30,000 Zulu warriors. When the Zulus got closer, they began to bang their spears on their cowhide shields and chant '*usuthu*' (kill). The disciplined but outnumbered redcoats bravely held their ground and prepared to fight to the death. There was nowhere to go – you could not outrun a Zulu *impi*. Nor should you listen to their terror-inducing death chants that formed part of the cacophony of war.

THE KILLING GAME

Believe it or not, farmers started it all – the killing game. When some of the very first cultivators of crops and livestock settled around the fertile Nile, Euphrates and Tigris Rivers, mankind stopped being nomadic. Holding onto land became a fight for survival. Early wars were over access to food and water, but as cities sprang up wars were fought between powerful city states, then unified kingdoms and mighty empires. Squabbling farmers were one thing, but organised warfare brought the violence to a whole new level.

A quick namecheck with the Assyrians, Babylonians, Egyptians, Hittites and Sumerians in the period from 3,500 BC highlights they are all famous for their military prowess as much as their civilising influences in the Middle East. The Sumerians were some of the very first people to conduct organised warfare and were the first to employ metal for weapons and wear armour. They were also the first to use the chariot and the phalanx. The destruction of Nineveh on the Tigris in 612 BC by the Babylonians marked the end of almost 2,000 years of Assyrian military dominance, leaving the Persians in the ascendancy.

Technologically, the evolution of warfare was painfully slow. During this long period cavalry and chariots were the shock troops of their day. They provided the mobile punch to terrorise enemy infantry. It is often referred to as the Age of the Chariot. Common foot soldiers were armed with the axe, sword, spear, shield and bow – notably the latter, although portrayed on ancient monuments from the end of the fourth millennium, was not used by charioteers until after 2000 BC.

From 500–323 BC, the ancient world was dominated by the turbulent Greek and Persian wars – proud city states against an empire, both bordering the Mediterranean. It was during this era that naval and siege warfare began to truly develop. Afterwards it was the turn of Carthage and Rome to fight it out for dominance. By this stage, the chariot had fallen from favour. Rome's legions and fleet eventually triumphed across the Mediterranean and Europe.

Following the collapse of the sprawling Roman Empire in AD 410, the so-called Dark Ages ran until 1066. Crucially, in all that time basic weapons remained largely unchanged, while numbers and superior tactics won the day. Then, in historical terms at least, things began to change very rapidly on the battlefield.

The second millennium AD of mankind was characterised by almost incessant warfare somewhere on the face of the globe. Sadly, in the field of human endeavour, it is one of the many things that the human race excels at. This book is not intended to glamorise or glorify war, but to chart and analyse its evolution on specific battlefields. It serves as a snapshot of the development of warfare over the last 1,000 years, illustrating the bravery and suffering mankind has inflicted upon itself in developing what we, rather crassly, call the 'art of war'.

In that time, warfare has progressed from the bow to the unmanned armed drone – staggering technological simplicity to staggering technological sophistication. The spear was replaced by the pike, which was made redundant by the musket, as was the bow. The musket evolved into the rifle, machine gun and sub-machine gun, which made the foot soldier the king of the battlefield. Catapults were replaced by cannon, which became field artillery, howitzers and anti-tank guns. Cavalry were replaced by the armoured car and the tank. At sea, the warship's reign was challenged by the submarine. Finally, warfare took to the air with the development of the fighter and bomber aircraft.

Increasingly, man has sought to dehumanise and depersonalise the 'art of war' by distancing himself from the physical business of killing his enemy, but one thing remains universal and that is the human condition. Much has been written about *esprit de corps*, but essentially the quality of troops and their leadership remains just as important, if not more so, as the quality or sophistication of their weapons. Excitement and peer pressure drive bravery, not loftier ideals of glory. Also, accurate intelligence has always been vital: where is your enemy, what is his strength and what are his intentions? Such information is essential in ensuring victory – get it wrong and it spells disaster.

Great acts of courage on the battlefield pose the question, what is it that pushes soldiers, sailors and airmen through the ages to feats of heroism? Why do they put themselves in harm's way? The ideal has always been for sovereign and country; deep down, though, it is acknowledged that servicemen and women invariably make the ultimate sacrifice simply to safeguard their comrades and families. Peer pressure is often a key factor when it comes to acts of selfless valour. Few go willingly to risk their lives or go to their deaths in the name of their country; however, risking ones' life to protect fellow human beings is a far nobler act.

Understanding the underlying psychological cause for heroism is one thing; it is another to fully comprehend how people individually or collectively muster the courage to face and often overcome life-threatening situations when all their instincts for self-preservation are urging them to flee. Remaining on any battlefield requires self-control and some courage. Good military training and discipline can only engender so much bravery when under fire. Ultimately though, humanity often shines out even in the face of the most brutal modern industrialised warfare.

In addition, there is an unspoken universal truth about war – some people enjoy it immensely because it is the most exciting thing they will ever experience. War is a drug and it can be very addictive. Nothing can replicate its intense buzz. Hernán Cortés conquered the New World for the thrill of it, as much as any divine mission. Soldiers fight because it is very dangerous and therefore very exciting.

While the basic nature of warfare has not changed over the millennia – namely defeating enemy forces and seizing ground – changing technology in terms of armour, naval and aerial warfare means that there have been fundamental changes, particularly in the past two centuries, in the types of fighting environment in which great courage is needed. Most notably, acts of gallantry by pilots, aircrew and submariners come to mind, where the elements themselves pose an equal, if not greater, danger to enemy action.

The following selection of campaigns, spanning 1066 to 2001, is entirely subjective and consists of engagements that have intrigued and inspired me as a writer in over thirty years of studying military affairs. The linking factor with each central battle is that they played a pivotal part in the outcome of the war. Their principal role is to graphically illustrate the changing face of warfare over the past millennium and how it shaped history. This ranges from the fierce Viking shield wall of the Dark Ages and the rise of the Norman knight, to the deadly long bow and knights of the Middle Ages, to the emergence of gunpowder during the Renaissance and pike-and-shot periods with the effect this had on siege warfare, to linear warfare in the Age of Enlightenment and finally to the rise of aerial and armoured warfare during the tail end of the bloodstained twentieth century. These are designed to give a flavour of the killing game and its far-reaching impact.

Novelist and former journalist Frederick Forsyth summed up the enduring fascination with the killing game when he wrote, 'A strange thing, war. With its bloodshed and cruelty, its pain, grief and tears, it ought to fill every civilised person with the utmost and unwavering revulsion. It ought to and often does. And yet, and yet …' Field Marshal Montgomery asked, 'Why do wars happen? Some will say that war is the child of civilisation, others that war stems from raw human nature. But war has always been the arbiter when other methods of reaching agreement have failed.'

1

THE VIKINGS ARE COMING: FULFORD GATE 1066

A great calamity befell northern Anglo-Saxon England in September 1066 when the country fought two major battles against marauding Viking forces. At Fulford Gate, the Saxon Army of the North was shattered and Norwegian King Harald Hardrada secured half of England. His Anglo-Saxon rival, King Harold Godwineson, managed to retrieve the situation at Stamford Bridge and snatch victory from disaster just before the Normans, under Duke William, invaded southern England.

While the battles at Stamford Bridge and Hastings remain poignant landmarks in British military history, few have heard of Fulford Gate or considered its significance. It was the very first engagement that fateful summer and Hardrada's triumph forced Harold to abandon his watch on the southern coast; ultimately, this dictated the relative weakness of his army at Hastings.

The only contemporary sources are Snorri Sturluson's thirteenth-century *Heimskringla* (*Saga of the Norse Kings*), in particular *King Harald's Saga*, and the *Anglo-Saxon Chronicle*.

King Harald's Saga is the most detailed account of the whole campaign, but its accuracy is open to endless debate. What is clear is that 1066 saw Anglo-Saxon England threatened with invasion from the south and the north. Both Duke William of Normandy and King Harald Hardrada of Norway had legitimate blood ties with the English throne. When King Edward the Confessor died that year, the Saxon Witan had favoured the succession of his opportunist brother-in-law, Earl Harold Godwineson of Wessex.

The seeds of the northern invasion were laid in 1065 when Harold's brother, Earl Tostig Godwineson, was expelled from his Northumbrian earldom. Harold did not spring to his defence, an act that could have plunged England into civil war. Tostig's rule had been so unpopular that it incited rebellion. To appease the rebels, King Edward I (the Confessor) had him banished and replaced him with the teenage Morcar of Mercia. Tostig claimed bitterly that Harold was implicated in the revolt – a slight possibility, as Harold may have seen him as a rival. The end result was that Tostig bore his brother a deep-seated grudge and was determined to regain Northumbria, no matter the cost. Furthermore, Tostig felt that he should have been crowned king on 6 January 1066 and not his brother. Initially, he raided southern England, possibly with the encouragement of Duke William, but was driven north until he found the sympathetic ear of King Harald Hardrada, Thunderbolt of the North.

Harald Hardrada was the last of the great Viking adventurers and had held a senior rank in the Byzantine emperor's Varangian Guard during 1035–44. His participation in the 1066 campaign heralded the end of the Viking era. Hardrada had a good claim to the English throne as he was related to England's Danish King Cnut (1016–35). Tostig may have offered his aid and possibly the co-operation of Northumbria in defeating Harold, if Hardrada would promise him the Northumbrian earldom or sub-kingship.

Tostig has always been cast as the villain of the story, which is not necessarily true; nonetheless, he had sailed first to Flanders and then Denmark seeking support. He hoped King Svein of Denmark would provide a Danish army, but ironically the king was too preoccupied defending Denmark against the Norwegians. So Tostig, legend has it, sailed to Norway to see King Harald Hardrada at Oslo Fjord. Hardrada was initially reluctant to invade England; there is some evidence to suggest that his son Magnus tried unsuccessfully in 1058 with a fleet from Norway, Ireland, Orkney and Shetland. In Norway, rather surprisingly, it was felt in some quarters that one Saxon *housecarl* was equal to two Norwegian warriors.

To the King of Norway, the opportunity of conquering fertile England was indeed a tempting prize. No doubt he felt confident that any Norman invasion would either not take place or could be contained and defeated in the South. His initial plans were simple – seize York, the capital of the North and the third city of the realm after London and Winchester. Finally, Tostig convinced Hardrada of the merits of the enterprise, and in the spring of 1066 Tostig sailed to Flanders to collect his English and Flemish troops.

By June–July 1066, the Norwegians began to gather their forces at Solund Isles and Hardrada sailed from Trondheim to collect them. Before leaving Trondheim, he took the precaution of having Magnus, his eldest son, declared king and regent in his absence. Even so, he took his wife and other children, including Prince Olaf. According to Sturluson, the king dreamt of his dead brother, who told him that death awaited him. Also, while on ship two Norwegian warriors had bad dreams bearing ill omens for the coming invasion – Gydir saw an ogress, who told him they were sailing west to die; while Thord, on a ship near the king's, saw an ogress riding a wolf prowling in front of the Saxon Army and consuming Norwegian corpses. In such superstitious times, these were foreboding portents of things to come.

By August, the fleet was on route for northern England. Fortunately for King Hardrada, the winds that blew down the North Sea aided his crossing and in turn blew across the English Channel causing Duke William to continually postpone his expedition.

King Hardrada had a very large army, although its exact composition can only be open to conjecture. He sailed from Sogne Fjord near Bergen with 200 longships and forty smaller vessels, which could have carried up to 18,000 men, a full *leidang*, and this was without the forces collected from the Scandinavian colonies off northern England. It is very unlikely that Hardrada would have taken Norway's entire fighting force. He probably only took the 7,000–8,000 professional soldiers that he is known to have had available (approximately 7,200 *hirdmen* fought at the Battle of Nissa in 1062 against the Danes).

England was not unprepared for the threat of invasion. The country was divided mainly between the three great earldoms of Wessex, in the South, and Mercia and Northumbria, which consisted of the Midlands and the North respectively. The rest of England was divided amongst Earl Waltheof, whose father had once been Earl of Northumbria (his earldom consisted of the shires of Huntingdon and Northampton), and Harold's brothers, Gyrth (Earl of East Anglia) and Leofwine (Earl of the shires of Bedford, Essex, Kent and Surrey).

Half of the war-making potential of England was centred on London and the other half on York. Indeed, the two northern earldoms combined had proved themselves a match for Wessex in 1050, when Harold's father was defeated

England in 1066.

in a power struggle. An Army of the North and an Army of the South (though Harold was probably only certain of the Norman attack) countered the threat of double invasion. King Harold, with the Saxon fleet, planned to protect southern England from the Normans, while Earls Morcar of Northumbria and Edwin of Mercia could defend northern England from Norwegian aspirations.

Anglo-Saxon organisation was based on the *fyrd*, or militia. Every freeman between 15 and 60 had a military obligation to serve in the General, or great *fyrd*, which was designed to meet local emergencies, but was badly trained and ill-armed. In contrast, the select *fyrd* was a more regular force, better equipped and prepared to fight outside their home areas. The majority of the select *fyrd* consisted of thanes – lesser nobles and the total national force may have been about 4,000. The population of England during this period was only between 1 and 2 million. Professional soldiers were provided from personal retinues of hearth-troops, or *hird*, very similar to the Vikings.

The royal household maintained a standing force of 4,000 *housecarls* (also spelled *huscarl*: the terms *hirdmen* and *housecarl* are largely interchangeable, although the latter came to cover all types of professional soldier), who were usually stationed near London and York under normal circumstances. Rather ironically, the Scandinavians provided mercenaries for the Anglo-Saxon armies and the Danes helped to found the original Saxon *housecarls* in about 1016. The Saxons' naval forces also provided fighting men in the form of *lithsmen* (sailors) and *butsecarls* (marines), some of whom might have been recruited into Edwin and Morcar's army.

The Norwegian *leidang* was a levy of ships and men, the nucleus of which was the *hird*. The *hirdmen* or *thingmen* paid retainers were organised under *jarls* (earls) who maintained sixty men, and supervised four *hersir* (local military commanders) who kept a further twenty men each. The numbers sixty and eighty being roughly two longships' complements. Commanding officers consisted of kings, sub-kings, princes and earls, while the senior *hird* officers were the *stallari* (marshal) and the *merkismadr* (marksman, standard-bearer). Hardrada had a personal *hird* of 120 men – sixty *hirdmen*, thirty *housecarls* and thirty *gestrs* (similar to select *fyrds*), commanded by *Stallari* Styrkar. The professionals were supplemented by freemen, peasants and *bondi* (land-owning farmers), although they were regarded as not particularly reliable.

Hardrada sailed first to the Viking kingdoms of Shetland and Orkney (he left his wife and daughters on Orkney) to gather the forces of Earls Paul and Erlend, Godred of Iceland, an unnamed Irish king, and the Faroes, the Hebrides and the Isle of Man, totalling perhaps 3,000. King Malcom of Scotland may have also unofficially contributed a few troops as he supported Tostig's cause. Sailing down the coast of Scotland the fleet numbered about 300 ships and up to 12,000 men, having been joined by a further 1,000 mercenaries and adventurers with Tostig. These forces, under Copsi, a rebel Northumbrian *thegn* and possibly Tostig's lieutenant, consisted of English and Danish *housecarls*, Flemish knights recruited in Flanders (the count was Tostig's brother-in-law) and Scottish mercenaries.

By 7 or 8 September, the Viking fleet had reached the Tyne Estuary. The two brothers, Edwin and Morcar, were caught off guard, having already called the *fyrd* out against Tostig earlier in the year. The Vikings terrorised the countryside for a week. Scarborough was burned and many of its inhabitants butchered for resisting, as were the local *fyrd*. Hardrada moved down the coast, landing at Holderness, where he engaged a Saxon militia force and easily defeated it. He then sailed southwards and by 18 September his fleet had entered the Humber Estuary, followed by the River Ouse. The Norwegians then disembarked at Riccall, a mere 9 miles from the prize of York.

News of the invasion may have reached Harold, back in London after his vigil on the southern coast, by about 15 September. He was now faced with the dilemma of whether he should march north and reinforce Edwin and Morcar or let them repel the invader on their own. The king

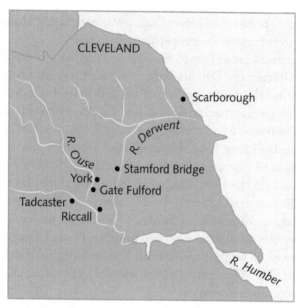

The area of campaign in 1066.

must have had a vague idea of the strength of the Normans across the Channel (about 8,000) but not the Norwegians, until they had actually landed. It transpired that the Viking forces were larger, and probably stronger than Harold had expected, therefore the invasion of the North was the more serious threat. Also, if the contrary winds in the Channel persisted there would be no Norman invasion before the end of the year and Harold and his supporters could rest easy.

Indeed, the situation in the North needed Harold's immediate attention. The inexperienced Morcar and Edwin were only in their late teens, with no reliable lieutenant to advise and support them. Coupled to this was Northumbria's Danish heritage, populated by a mixture of Saxons, Danes, Norse and Irish Vikings – there was a slight risk that they might rise up in support of Hardrada or rally to the discredited Tostig. Harold must have seen it as a good chance to combine the Southern and Northern armies in order to deliver a knockout blow to the Norwegians. This would be good for the country's morale and would set an unsettling example to Duke William. If he was successful, he could then march south with a very large army, in the sure knowledge that York was secure. On about 18 September, the king marched northwards, gathering 3,000 men, most of whom were regular soldiers, *thanes*, *thegns*, *housecarls* and select *fyrd*.

Meanwhile, in the North, the two earls had remained questionably inactive. Why had they not attempted to confront the Viking invasion force on the coast? To start with, Northumbria was not as populated as the South so it would have taken longer to gather their scattered forces. During the week that Harald Hardrada had been in England, the earls had been hurriedly gathering the select *fyrd*, which were probably quite weak and mustering the northern great *fyrd*. Also during this period, the Norwegians had been effectively drawn inland, stretching their lines of communication and dividing their army in order to protect the fleet. It seems the Saxons may have had a weak naval force at Tadcaster on the River Wharfe, this meant if the Vikings ventured beyond the junction of the Rivers Ouse and Wharfe there was a danger of them being cut off from the sea. This may account for why Hardrada landed at Riccall instead of further up the Ouse.

Even so, by 19 September Edwin and Morcar must have been very alarmed by how close the Norsemen were to York. The city was reasonably well fortified, although its population, numbering some 9,000, must have been swelled by refugees fleeing the marauding Vikings, plus the earls had gathered an army in the vicinity. It is very doubtful whether the city could have sustained such a large number of people for long, or withstood a siege. Fortunately, the Norwegians were not equipped for a siege, but nonetheless Edwin and Morcar could not stand by and do nothing while Hardrada's forces continued to ravage the North. They could not really afford to wait for Harold, even if they had heard he was on his way, which is doubtful. The king had over 200 miles to cover and they probably did not expect him for at least two weeks. Edwin may have heard that Harold had fallen ill (legend has it) on his return to London and did not expect him at all in the immediate future.

The two earls were probably spoiling for a good fight as neither had fought in a major battle before. They had gathered a reasonably large army and had been joined by Earl Waltheof. He was only about 20 and likewise may have been keen to get to grips with the invader. The three earls, on deciding to give battle, probably had little grasp of the strategic implications of their actions. They marched their army to Fulford Gate (also Gate Fulford) or *Apud Fulford* (Fulford Water), 2 miles south of York to block the Norwegians' line of advance on 20 September.

On that very day, King Hardrada marched on the city, leaving Earls Paul of Shetland and Erlend of Orkney with about 2,000 men to guard the fleet, and possibly up to 1,000 men in Cleveland, Scarborough and Holderness. There was only one road leading to York, up which the Viking host would have to force their way. It was raised because of the nature of the surrounding countryside; several hundred yards to the west lay the River Ouse, while to the east lay a water-filled ditch and beyond, a marshy fen.

The only reasonably contemporary source, Snorri Sturluson's *Heimskringla*, says the Saxons had 'an immense army', which is probably true. It may also partly account for why the Saxons were so ready to give battle. Morcar's Army of the North consisted of his own *housecarls*, plus the

Northumbrian select *fyrd* and the poorly equipped great *fyrd*. The estimated select *fyrds* of Mercia and Northumbria were about 8,000 men; this would have given Morcar 3,000–4,000 select *fyrd*, along with say about 2,000–3,000 great *fyrd*.

Edwin had arrived with his *housecarls* and the Mercian select *fyrd*, probably numbering in the region of at the most 2,000 fighters, while Waltheof's contingent was likely to have been no stronger than 1,000. The total number of *housecarls* with the army may have been about 750, based on the personal retinues of the three earls – roughly 250 men each. In 1065, Tostig is known to have had 300 *housecarls* in his service. It is quite possible that there may have been up to 1,500 royal *housecarls* at Slessvik near York, but it seems more likely that the vast majority of them were in the South with Harold.

Therefore, it appears the Saxons had an army numbering in the region of 8,000–9,000 men, roughly the equivalent strength of the Vikings, but they had one big disadvantage, they were largely a provincial army being pitted against a largely professional fighting force. Nevertheless, Northumbrian confidence in their army was strengthened by the fact that 100 clergymen from York marched to Fulford Gate with them.

On Wednesday, 20 September 1066, the Saxon Army of the North finally confronted the Viking Army of Norway to settle the first round of the fateful struggle for the English crown. In order to block the Vikings' line of advance, the Saxons straddled the road, their line stretching from the River Ouse to the marshy ground towards Heslington. They deployed in a close formation divided into two or three divisions. Morcar commanded the Saxon left and Edwin the right, with Waltheof in the centre, or more likely on the far right. (This is justified by his subsequent successful flight from the battlefield.)

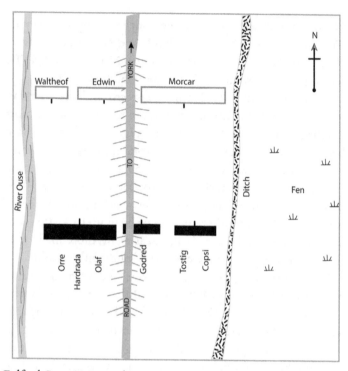

Fulford Gate, 20 September 1066.

By the eleventh century, the Saxons were using horses not only for transport (*housecarls* often rode while on campaign), but on limited occasions as weapons of war. It was no doubt a natural progression influenced by developments on the continent. Even so, in all their major engagements the Saxons fought like their Scandinavian cousins – on foot. Snorri Sturluson's account of Saxon cavalry at Stamford Bridge possibly refers to mounted *housecarls*, but may be confused with Hastings and is therefore potentially misleading. Certainly, no Saxon cavalry are mentioned at Fulford Gate. The Saxon battle formation was the *bord-weal* or *scyld-burh*, alias the shield wall, and tactics were very similar to the Vikings.

Scandinavian armies fought predominantly on foot, although they did use horses for transport and raiding. Viking tactics consisted of the shield wall and the *svynfylking*, a wedge-shaped formation, designed to pierce an opponent's line. They fought about five deep and were sometimes broken up into divisions, as seems to have been the case at Fulford Gate. Sword and axemen formed the front rank, then the spearmen, followed by the archers, javelin men and slingers. A battle commenced with a hail of missiles to weaken the opponents' shield wall.

How calculated Hardrada's deployment was, is unclear, but it was to have significant bearing on the forthcoming battle. Hardrada, seeing the Saxon Army arrayed before him, placed the bulk of his Norwegians on the left, next to the river. He personally took command of this wing with his son, Prince Olaf, and his lieutenant, *Hird Stallari* Eystein Orre. Godred of Iceland may have commanded the centre, while Tostig, possibly supported by Copsi, commanded the right. The centre and right stretched thinly down to the ditch, with the left probably twice as strong.

Ironically, it seems the opposing leaders deployed their forces so that their weakest wings were facing their opponents' strongest one. Morcar was on the Saxon left with the bulk of the Northumbrian forces, while Hardrada was on the Viking left with the majority of his Norwegian troops. The battle that followed can be divided into three distinct phases.

Morcar, eyeing the Viking line, could see Hardrada's battle standard on one wing and Tostig's on the other. The latter's flank was visibly the weaker; this would be the point at which to turn the Norwegians. In the first phase, the Saxon Army marched forward, Morcar with his Northumbrians advanced down the line of the ditch, leaving his brother and Waltheof to hold Hardrada. Upon reaching Tostig's shield wall, the Northumbrians let loose a hail of missiles and then with battle cries of 'God Almighty' and 'Holy Cross', they crashed into Tostig. Under the pressure of the impact, the Danes, English, Flemings and Scots began to give ground – this may have been a deliberate move – and the Northumbrians pressed forward. Morcar and his *housecarls*, hewing about them, tried to cut their way to Tostig but his line did not break.

At this point Harald Hardrada could see that the Saxon left had outpaced the right. With a cry to his *merkismadr*, brandishing his standard *Landeythan* (Land-Waster or -Ravager) and surrounded by Styrkar's *housecarls*, he led forward his fierce Norwegians. In this second phase, Hardrada's shield wall charged into Edwin's: so great was the fury of the Vikings' assault that they swept away all before them. They pierced the Saxons' centre and shattered their right. Edwin's banner went down and the Mercian line collapsed. Throwing away their arms, his men fled upriver, into it, or towards Morcar in panic. Unfortunately, Morcar's plan had misfired and it was the Saxons' flank that had been turned, not the Vikings'. Possibly they had fallen into a deliberate trap.

In the final phase of the battle, Hardrada swung his forces to the east, pursuing the broken Mercians and cutting off Morcar's remaining men. Tostig also urged his warriors forward and Morcar's troops were pushed back, being trapped against the ditch and the fen. For the Saxon Army of the North the final collapse was almost over. Many Northumbrians and Mercians fled across the fen and were drowned; others were cut down by the victorious Vikings.

Morcar's *housecarls* died fighting around the young earl, while Edwin's men were pursued to the very gates of York where many more were cut down. The Saxons left the ditch, according to Sturluson, 'so filled with dead that the Norsemen could go dry-foot over the fen'. About 1,000 of the North's best troops were left strewn over the battlefield, probably mainly *housecarls*, along with the clergymen from

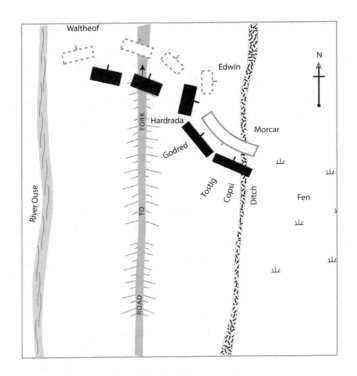

Collapse and destruction of the Saxon Army.

York who were also mercilessly slaughtered. The *Anglo-Saxon Chronicle* states the Northumbrians made 'great slaughter' of the Vikings, so it would be safe to assume the Norwegians lost about 400–600 men.

Sturluson gives the impression that Morcar was killed, 'Warriors lay thickly fallen, around young Earl Morcar', but this simply means his *housecarls* died protecting him. Waltheof managed to flee with the survivors, but it is quite likely Edwin and Morcar were captured, as they would have been valuable for ransom and as hostages. Certainly, all three earls were to later lead northern resistance to the Norman occupation in 1069.

The military power of Mercia and Northumbria was scattered. York, fearing a repeat of Scarborough, surrendered, offering its allegiance, men and hostages. Hardrada demanded 500 hostages, soldiers for the invasion of southern England and then withdrew to Riccall. King Harold had left London on the very day of the battle and arrived in Tadcaster, 10 miles from York, on Sunday, 24 September having already heard of the earls' defeat. He must have realised they had jumped the gun and he would have to confront the Norwegians largely unassisted. At Tadcaster, Harold drew his army up in battle order expecting an attack from York, when none came he marched into the city the following morning. Harold was faced with three courses of action: he could stay in York and await attack, march on Riccall or advance to Stamford Bridge, where the Vikings were to collect their hostages, and try to catch them by surprise. Harold's men rested briefly and then marched on Stamford Bridge.

After his victory at Fulford Gate, Hardrada seems to have thrown caution to the wind, assuming that the North would not offer any more serious opposition. That Monday, he disembarked his army and once again divided his forces as he had done prior to Fulford Gate. From each company, two men were to go for every one that was left behind. Prince Olaf, *Stallari* Eystein Orre and Earls Paul and Erlend also remained with the fleet. Therefore, Hardrada probably took with him 7,000–8,000. Tostig accompanied him, as did *Stallari* Styrkar, commanding his personal *housecarls*, possibly Godred of Iceland and Copsi.

It was a hot day, so Hardrada's men foolishly discarded their mail and leather armour, taking only helmets, shields, swords, axes, spears and bows. After all, they were marching to a rendezvous with a defeated people, not a battle – or so they thought. Stamford Bridge, over the River Derwent, is

7 miles east of York, all the roads in eastern Yorkshire converged here and it strategically dominated the region. On arrival, the Vikings casually strew themselves in the grass on either side of the river, perhaps recounting their exploits at Fulford five days previous. Harold was presented with the opportunity of defeating the Norwegians piecemeal, it is possible he knew of the 3,000 men still at Riccall, but even if he did not, the army before him was unprepared, divided and not in a very favourable position.

The Saxons had, in the meantime, marched along the old Roman road to Gate Helmsey. It was only when they were a mile away, coming over the brow of the gradual slope from Gate Helmsey to Stamford Bridge that the Norwegians spotted their banners. Hardrada summoned Tostig, who thought they looked hostile but might be kinsmen coming to join them, so there was a vague possibility that they might be rebels. The Norwegian King waited cautiously. Sturluson notes, 'the closer the army came, the greater it grew, and their glittering weapons sparkled like a field of broken ice.'

When the Golden Warrior standard came into view Hardrada rapidly realised it was King Harold Godwineson. Tostig counselled retreat to the fleet and the rest of the army, but the majority of the Vikings were on the wrong side of the river to be able to retire to Riccall quickly. Most of the Norwegians were on the left bank and to retire down the right bank would probably require a major rearguard action as the Saxons were so near.

King Harold had marched north with about 3,000 men; many of them would have been professional soldiers. Once in the North, he was probably able to recruit up to 2,000 more, but it seems doubtful that the bulk of Morcar's and Edwin's scattered army was retrievable. This means that Harold had a force of about 5,000, possibly 6,000 at most, which were visibly outnumbered by the Vikings. Hardrada decided to send three men on horses to fetch the others while he would fight a holding action. Hardrada would not have liked the idea of fighting with his back to the largely unfordable river. However, the slope on the right bank immediately above the bridge, in an area now called Battle Flats, was an attractive defensive position. Therefore, those Vikings on the left bank were ordered to delay the Saxons while the main body formed up on the right side of the river.

Sturluson talks of cavalry attacks, and it is most likely at this point that one occurred. The mounted Saxon *housecarls*, on seeing the Vikings divided, spurred their horses towards the bridge. Some Norwegians withdrew, but others formed a semi-circular shield wall blocking the Saxons' way. Many of them must have dismounted to get to grips with their foe, who, under the weight of numbers gave ground and eventually collapsed. One solitary Viking, wearing a mail coat and wielding a two-handed axe, prevented the Saxons from clinching victory by straddling the bridge and holding everyone at bay or chopping them down. An arrow was shot at him but glanced off his armour. Anyone who tried to rush him was hewn down; it is claimed some forty men suffered this fate. Finally, in a rather cowardly act, someone in a boat stabbed him from underneath the bridge and the Saxon Army streamed across to find the Norwegian Army formed up on the slight rise.

Hardrada did not attempt to further prevent the Saxons from crossing because this would have simply brought the battle to a halt. He drew his army up in a circular shield wall, placing himself, his personal *housecarls* and *Merkismadr* Fridrek with his black raven standard, Land-Waster, in the centre. Both Hardrada's *hirdmen* and Tostig's men were to act as the reserve. The front rank formed a spear wall by embedding

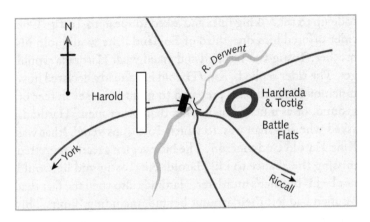

Stamford Bridge, 25 September 1066.

the ends of their spears into the ground, while those behind were to thrust their spears at the horses' chests. This was done, according to Snorri, because the enemy cavalry always attacked in small groups and then wheeled away.

All this sounds very uncharacteristic of Viking or Anglo-Saxon warfare. However, because of his Byzantine experiences, Hardrada would have been familiar with the Byzantines' *kontos* 12ft spear or the *rhiptarion* 9ft spear. When attacked by cavalry, Byzantine infantry's front rank would brace themselves by pushing their spear butts into the ground. Thus, Tostig's Flemings may have been some of the first people in western Europe to use the pike under Hardrada's guidance. Also, some of the Norwegians may have used kite shields, although the Varangian Guard is not recorded as using them until 1122.

King Hardrada, while inspecting his warriors, fell from his stumbling horse and Harold, upon seeing this, was awe-struck by his size. Shortly after this, twenty Saxon *housecarls*, all in mail (including their horses according to Sturluson), rode up to the Viking lines and asked to speak to Tostig. One rider offered him one-third of England if he would join his brother. Tostig was insulted and asked what Hardrada would get. The rider replied, 'King Harold has already declared how much of England he is prepared to grant him: seven feet of ground, or as much as he is taller than other men.' Hardrada asked who this man was, to which Tostig answered, 'that was King Harold Godwineson'. The Norwegian greatly regretted missing the chance to kill Harold, but Tostig said he would not be his brother's murderer. Hardrada also rued the fact that his men had left their armour behind, including 'Emma', his knee-length mail coat.

Snorri Sturluson says the Saxon cavalry now attacked, riding round the Norwegian shield wall. They, in return, fired their bows at the horsemen. The Saxons repeatedly charged, regrouped and charged again. At one point the Vikings pursued the retiring horsemen, but the Saxons turned and rode them down (this all sounds too much like Hastings …). King Hardrada was enraged and charged to help his men and in full battle fury all gave before him. Many were slain and it looked as if the Saxons would be routed. At this crucial moment, Hardrada was struck in the throat by an arrow. He fell, and all those around him were slain, although some retreated under Land-Waster. Indeed, Tostig's forces rallied under this banner and both sides fell back to regroup.

Harold offered the survivors quarter, including his brother, but all refused, preferring to die with their king, and the battle resumed. It was about now that Eystein Orre belatedly arrived with his 3,000 reinforcements from the fleet. They had forced marched the 12 miles from Riccall, crossing the Derwent to the south at Kexby.

Reaching Tostig's position, Orre seized Land-Waster and his forces fell on the Saxons who, at this new onslaught, again nearly broke. However, Orre's men had run all the way from the fleet in their armour and most were exhausted. As the fighting progressed, many of them threw off their mail; some collapsed, and others even died of exhaustion. The result of this was that the force drained out of Orre's attack and the Vikings were finally beaten. By the late afternoon it was all over; some remained to fight to the death, while others fled.

Stallari Styrkar, commander of Hardrada's personal *house-carls*, managed to escape on a horse; Godred of Iceland also fled and escaped to the Isle of Man. In the Lincolnshire village of Barrow-on-Humber, a legend persisted that after the battle Hardrada lived there as a hermit, and although this is untrue, it may indicate that some Vikings fled there. The defeated Norwegians were pursued back to their fleet, where the fight continued. As the *Anglo-Saxon Chronicle* notes, some of them were drowned, others burned, which may indicate that some of the ships were set alight.

King Harold gave quarter to the survivors, including Olaf and the Earls of Orkney. They were sent home in just twenty-four of the original 300 vessels. The Vikings are thought to have lost as many as 7,000 men, although this may be a little high, and many of them would have been butchered in the surrounding countryside. Tostig's body was found on the battlefield with his head cloven right to the chin by an axe blow, possibly delivered by Harold himself. Hardrada lay surrounded by those he had slain.

It was a whole generation before the Norwegians undertook another foreign expedition. Stamford Bridge was the last major battle fought on English soil in which a Viking Army took part. Three years later, the Danes arrived with a fleet to join the Saxon rebels resisting the Normans, but they were bought off.

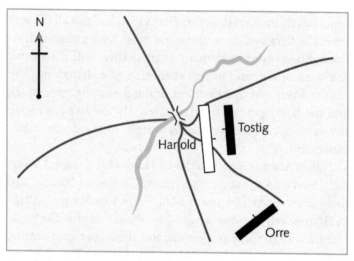

Stamford Bridge, the final battle.

King Harold's victory celebrations were cut short on 1 October 1066, when he heard that Duke William had landed on the south coast at Pevensey three days earlier on 28 September. By winning at Stamford Bridge, Harold's victory became the all-important decisive Battle of the North, pushing Fulford Gate into the shadows of obscurity. The significance of Fulford Gate is clear, if the Norwegians had been defeated earlier, a far stronger Saxon Army containing a large northern element could have confronted the Normans at Hastings and the outcome of the day could have been completely different. Instead, a fortnight later, Harold was dead on the field of Hastings and William was on the military road to becoming master of England. By retrieving the disaster of Fulford Gate with victory at Stamford Bridge, Harold tragically undermined his chances with William, though it was to be a close-run thing.

2

CHIVALRY IN ARMOUR:
NORTHALLERTON 1138

In the wake of the relative peace and unity under William the Conqueror and his sons, England lapsed into civil war and anarchy for nineteen bitter years. In fact, the situation was so bad that it has been compared to the Wars of the Roses and the English Civil War, which occurred hundreds of years later. One of the root causes was that during the late eleventh and early twelfth centuries rapid castle building rendered major pitched battles almost superfluous. Notably, while this nineteen-year period witnessed only two major military engagements, at Northallerton and Lincoln, there were numerous protracted sieges. The most unusual of the two battles, Northallerton, saw an English Army led by a holy standard dramatically drive off a Scottish invasion supporting a contender to the English throne.

When William II died in 1100, his younger brother Henry I, who has been described as a grasping opportunist, succeeded him. Certainly, after William's death in a dubious hunting accident, Henry was quick to seize the royal treasury in Winchester and move to Westminster with his supporters.

This swift action pre-empted his older brother Robert, who was away on the First Crusade. However, Henry's only legitimate son William was killed in a shipwreck in 1112 and his second marriage did not provide another male heir.

It was Henry's desire to see his bloodline continue that laid the seeds of civil war. In December 1126 he gained an oath of fealty from the barons supporting his daughter Matilda's accession. She was to prove unpopular because of her eventual marriage to Geoffrey of Anjou, the French Angevins being the traditional enemy of Normandy. From 1114 Matilda had been married to Emperor Henry V of Germany and remained empress until his death in 1125. Three years later, she married Geoffrey of Anjou to complete a new military alliance.

There were actually three other contenders for the English throne. The candidates were Henry's illegitimate but able son, Robert, Earl of Gloucester, who had seen military service against French and Norman rebels, Henry's favourite nephew, Stephen of Blois, and Stephen's elder brother, Theobald of Blois. In particular Stephen, son of Stephen, Count of Blois and Adela, the Conqueror's daughter, coveted the English throne. He had no English endowment and was landless until he astutely married Matilda (Maud), daughter and heiress of Eustace, Count of Boulogne in 1125, making him the second richest baron in England and Normandy. To complicate matters, Maud's mother, Mary, was sister of King David I of Scotland and Matilda the Good, first wife of Henry I. In fact, Maud may have been the motivating power behind Stephen, urging him to act after Henry I's death. A year after Stephen's marriage, his younger brother, Henry of Blois, became Abbot of Glastonbury and in 1129 also became Bishop of Winchester. Until his death Henry of Blois held the richest abbey and the richest see in the country, making him a very powerful man.

Upon King Henry I's death in Normandy in 1135, Stephen moved quickly to take the throne. Arriving from Boulogne he bypassed Dover and Canterbury which were shut to him. In London, he gained the support of Geoffrey de Mandeville, Castellan of the Tower of London and then moved to Winchester to secure the government. With the help of his brother, the Bishop of Winchester, he gained the royal treasury, some £100,000, as well as considerable plate and jewellery. He would need this money to finance the coming wars.

Hugh Bigod, steward of the royal household, was bribed to inform the Archbishop of Canterbury that King Henry had disinherited his daughter on his deathbed. Through Henry of Blois the support of the clergy was guaranteed and the majority of barons initially declared for Stephen. Even Pope Innocent II supported his accession. Thus persuaded, the Archbishop of Canterbury crowned Stephen on 26 December 1135.

To secure his position Stephen granted royal rights, titles and lands, but after taking the throne in such a decisive manner, he was to prove a weak king who was neither able to subdue his enemies nor control his allies. He was brave in battle, but this was not enough. One contemporary chronicler described Stephen thus, 'of outstanding skill in arms, but in other things almost an idiot, except that he was more inclined towards evil'.

It was now that the flaws in the feudal system imposed by William the Conqueror began to show, for, without the complete loyalty of the barons, a king was nothing. Stephen's reign was to be the most anarchic England has ever seen and has been described as 'the nineteen long winters when God and his saints slept'. The barons soon looked to themselves, the Exchequer ceased to operate and the sheriffs stopped collecting royal revenues.

Abroad, Empress Matilda and Geoffrey of Anjou agreed to divide Stephen's possessions, with the Angevins taking Normandy and Matilda England. She had powerful allies who would support her claim: as well as Robert of Gloucester, her half-brother, King David I of Scotland was her maternal uncle. Through his wife's family, David laid claim to Northumbria, Cumbria and the earldom of Northampton and Huntingdon. He had everything to gain from supporting the empress.

King David acted swiftly, invading northern England in the winter of 1135, occupying the key border strongholds of Carlisle, Wark, Alnwick, Norham and Newcastle. The castles of Wark and Helmsley belonged to Walter Espec, Sheriff of Yorkshire and a royal justice, so he was quick to call for help, but King David's march south was only stopped by the arrival of Stephen at the head of an army in February 1136. The powerful Robert of Gloucester had paid homage to the new king, probably persuading Stephen that his position was moderately secure in the south.

Stephen's policy towards the Scots was one of appeasement. An agreement was signed to the effect that David's son, Prince Henry of Scotland, should take possession of the castles of Doncaster and Carlisle. Also, he should be recognised as the Earl of Huntingdon and Henry's claim to Northumbria should be recognised in favour of his stepbrother's (Simon de Senlis' son, by David's wife's first husband). Scotland's southwestern boundary was now fixed at Carlisle and the River Eden. Unfortunately, Stephen's negotiations while avoiding battle simply convinced David that the English King was a weakling who could be scorned. Ironically, Stephen felt confident that Scottish influence had ended sufficiently to allow the newly fortified castle and town of Carlisle to mint new coins in 1136, possibly to help finance the garrison.

Back in the south, two separate rebellions broke out. Hugh Bigod, unrewarded for his perjury, took the opportunity to take possession of Norwich Castle, while on rumour of Stephen's death, one of Matilda's supporters, Baldwin of Redvers, took the opportunity to pillage Exeter and occupy the castle.

Robert of Gloucester, operating from his power base at Bristol, supported Stephen's campaign against Exeter Castle in 1136. Royal forces, however, refused to prosecute the siege with any vigour. Only after three months did the garrison, now surviving on wine as the well had dried up, finally surrender. Instead of making an example of the rebels as was customary, Stephen let them go unpunished. The message was clear – Stephen the merciful could be taken advantage of.

Across the Channel, when Geoffrey of Anjou moved against Normandy, the Normans called on Stephen's elder brother, Theobald of Blois. The Norman duchy also declared its allegiance to the English crown. Stephen sailed to the continent in 1137 to secure control of his French domains but the campaign against Geoffrey of Anjou proved to be a complete fiasco. Stephen's Flemish mercenaries, under the brutal William of Ypres, antagonised their Norman forces to such an extent that his army disintegrated. This again illustrated what a weak leader Stephen was and swallowed up the last of Henry I's much depleted treasure. The end result was that Normandy was not protected against the Angevin threat and disastrously Stephen could no longer rely on Norman military assistance.

The troublesome Scots, in the meantime, invaded again that year and despite a truce returned once more in January 1138. David was determined to enforce Stephen's concessions. King Stephen led an army north and ravaged the lowlands to overawe the Scots – at least this was an act his enemies could understand. Then his army, loaded with plunder, shrugged off his control and drifted home. David moved on Newcastle and his nephew William defeated Stephen's blocking force at Clitheroe in Lancashire.

In light of Stephen's poor military control, some of the northern barons decided it was time to side with David. During the summer of 1138 the English Baron Eustace Fitz John, with his key castles of Alnwick and Malton, defected. In particular, Alnwick had always served as a vital strategic bulwark against Scottish invasion. Crossing the Tweed, the Scots, up to 22,000 strong and described as a 'barbarous multitude', invaded England in the summer of 1138. David was serving his own interests as well as Matilda's, for he still wanted to get back the shires of Huntington and Northampton, which had been taken during William the Conqueror's reign.

At the same time, Bristol was in revolt under Robert of Gloucester, with the support of the Constable of Dover. Robert had been alienated by his rivals' support for the new king and once the Angevin threat had been partially thwarted in 1138–39, he threw his lot in with Empress Matilda. This action ensured that Stephen was tied down in the south. He dared not return north because of the two-pronged threat posed to London and Winchester by Dover and Bristol respectively. Indeed, during 1138 it was his wife, Queen Maud, who oversaw the capture of Dover. All Stephen could do was send Bernard of Balliol with a force of household knights.

The Scots plundered Northumberland and north Yorkshire with such violence that their behaviour was likened to wild beasts, and in Hexham the women were bound and sent back to Scotland as slaves. The man of the moment was Thurstan, the elderly, invalid Archbishop of York. He also held the post of Lieutenant of the North, and as the king's appointed deputy it was his task to defend the north of England. Thurstan took the novel approach of declaring a

Holy War, guaranteeing a flock of volunteers seeking remission for their sins. David had allowed the men of Galloway with their wonton massacres to alienate the allied northern barons, driving them back to Stephen's cause. They answered Thurstan's call, including Walter Espec of Helmsley, Sheriff of Yorkshire, William Earl of Albermarle, Robert de Ferrers, William le Gros, Gilbert de Lacy, Roger de Mowbray, William Piercy, de Courcy, Bruce and Balliol. They gathered an army perhaps 10,000 strong.

The Norman feudal military levy now supplemented the old Anglo-Saxon *fyrd*. On the basis of the Domesday Book, the magnates providing military service to the king were not significant. William the Conqueror relied on less than 200 barons, who supplied a core of at least 4,000 fully equipped knights. The English Church's military obligations provided another 780 knights, so including peasant levies English armies of this period would have amounted to little more than 15,000 men. Under Stephen the system was weakened by the constantly warring factions and those nobles remaining aloof and seeking to preserve their military resources.

William had relied on kinsmen and trustworthy lieutenants to hold the key military commands, but by the 1100s that bond of loyalty had been eroded. Originally these men had been granted land in return for knight service, although increasingly scutage (tax) allowed those with military obligations to provide money instead, which was used to raise mercenaries. Indeed, Breton, Flemish and Norman mercenaries were to become the scourge of the kingdom. Thus, knight service was subdivided, dissipating the system even more. In many ways the mercenaries were preferable, for under the feudal system a man only had to serve forty days, while hired troops could be contracted for much greater periods. Ironically, Stephen's troublesome mercenary captain,

William of Ypres, proved to be more loyal than many of his baronial vassals.

To give his army religious inspiration, Thurstan produced a holy standard. A mast was fastened to a four-wheeled cart, topped with a sliver pyx containing a consecrated wafer, while from the two arms of the cross dangled the banners of St Peter of York, St John of Beverley, St Wilfrid of Ripon and St Cuthbert of Durham. Unable to don his armour or march with the army because of illness, Thurstan sent Ranulph, Bishop of Durham, as his representative in the field. They moved north as far as Thirsk, where Balliol and Bruce rode forth to meet with King David and offer him the earldom of Northumberland, if he would withdraw his forces. In the face of David's rejection, Balliol and Bruce renounced their own fealty to the King of Scotland.

Thurstan's holy army then marched through Elfer-tun, now called Northallerton, to Cowton Moor, 3 miles to the north, halfway between York and Durham. Here, there were two hillocks 550m apart, and the English positioned them-selves on the southern one with the standard wagon placed on the summit. The army consisted of three divisions: the front comprised bowmen from Yorkshire and Lincolnshire, the second spearmen and the third, gathered round the standard, knights and men-at-arms. The knights had sworn to defend the sacred symbol to the death. Shire levies may also have been placed on each wing and to the rear. The English knights were to fight on foot, sending their horses to the rear with the baggage train near Northallerton Castle in Scotspit Lane. A small detachment of horsemen was deployed to guard them.

The Scots, numbering 10,000–15,000 with possibly about 1,000 English and French supporters, occupied the northern hillock. David placed his knights, men-at-arms and archers in the vanguard, with the Galwegians and Highlanders behind.

These dispositions were disputed amongst the Scots leadership. The men of Galloway, under the Earl of Lothian, insisted they should lead the attack, and after a row between the Earl of Strathearn and Alan of Percy, David foolishly acquiesced. The Galwegians under Lothian, Donald and Ulgerich now formed the centre and the front line. Traditionally fighting nearly naked, these men were lightly armed with long pikes or javelins and were to prove a liability.

The right wing, slightly behind the centre, was made up of the men from Strathclyde and the eastern lowlands commanded by Prince Henry. He had most of the mounted knights and his standard was, appropriately, a lance with a bunch of flowering heather. The West Highlanders, the men of Argyle, the Hebrides and Lorn, with the Lothian and English knights, were on the left, led by Alan of Percy. David was either behind the centre or left with a small reserve comprising a bodyguard of mailed knights and the men from the East Highlands and Moray.

Battle commenced in the early hours of 22 August 1138. The Bishop of Durham said a prayer beneath the standard, with the whole of the English Army kneeling to receive absolution. Espec, a second generation Anglo-Norman, may also dramatically have made a rousing speech recalling the traditions of Norman valour to inspire the men. As Sheriff of Yorkshire, Espec might have been in command, but he may have deferred to Balliol as the king's representative.

The bowmen of Yorkshire and Lincolnshire met the unarmoured Galwegians' opening attack with predictable and bloody effect. According to the Abbot of Rievaulx, who was present, the Galwegians looked 'like a hedgehog with quills'. Ironically, over 200 years later the French were to suffer a similar fate at Crécy (see Chapter Four). Nonetheless, the Galwegians seem to have penetrated the English front rank

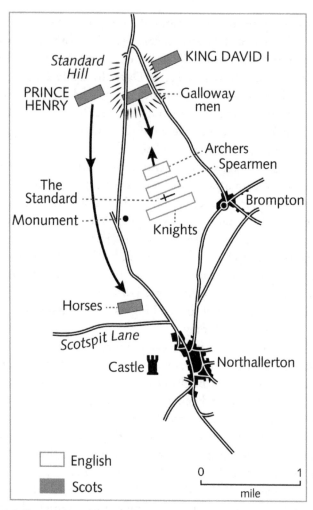

Northallerton, 22 August 1138.

with sheer momentum, though their success was short lived. With the war cry of '*Albanach*!' upon their lips they made a second and third charge. The archers rallied to continue their flights of arrows and Lothian was killed. His men fell back, becoming entangled in the second line. The shafts continued to fall when the English smashed the front of the Scots centre, and the third line containing the English and French held.

Prince Henry, hoping to end the stalemate, charged the English left flank with about sixty horsemen. They did this with such force that he reached the horses and baggage train. Unperturbed, the English simply closed ranks after him to face the Strathclyde foot soldiers. Now outnumbered, Henry found himself in danger of being trapped by the English third division. In the confusion his men, throwing away any field signs and insignia, merged with the English knights before effecting their escape.

The Strathclyde forces were pushed back and the remaining Galloway men, with Donald and Ulgerich slain, began to stream from the field. The as-yet-uncommitted left wing of the Scots Army, who could have saved their centre, launched one half-hearted attack only to be beaten off, after which they withdrew. The English centre joined the battle on the wings overwhelming the Scots.

A charge by the English knights finished the matter. David tried to commit the Scots reserve, but his advance melted away until all that was left were his bodyguard. Seeing the day was lost, he retired to a nearby hill to rally his remaining forces. They were numerous enough to deter the English from resuming their attack, and in the chaos King David and Prince Henry managed to flee separately towards Carlisle. It appears the English Army did not give chase, although the Scots retreated as far as the Tyne.

The battle lasted two hours and up to 10,000 men were killed, of whom perhaps only about 100 were English, including de Lacy. Some fifty Scottish knights were captured, possibly from Henry's charge. In terms of Scottish defeats, the Battle of Northallerton, or 'the Standard', as it is sometimes called, is comparable to Alnwick in 1093, Neville's Cross in 1346 and Flodden in 1513. Once victory had been achieved, the English Army dispersed, leaving a force to besiege the rebel Eustace Fitz John's Malton Castle.

Just three days after, David rallied his forces at Carlisle, where he was joined by his son. He then laid siege to Wark Castle, which fell to him. To avoid further Scottish invasions, Stephen foolishly recognised Henry's claim to Northumbria under the Treaty of Carlisle. Ironically, although defeated David still advanced his frontier in the south-east to the River Tees, with the west already secured on the Eden. Cumberland, Westmoreland and Northumberland remained, for many years, part of the kingdom of Scotland.

Despite thwarting the Scots, Stephen was facing full civil war at home. Having secured his northern borders, he was forced to turn south to deal with the troublesome Empress Matilda. Many of the barons of the West Country and the Northern Marches had declared for her. Fearing the clergy and coveting their wealth, Stephen arrested Roger, Bishop of Salisbury, along with his nephews, Alexander, Bishop of Lincoln (who was believed to be Roger's son) and Nigel, Bishop of Ely. The latter escaped to Devizes Castle. The king's response was that Roger and Alexander would starve until Nigel surrendered himself, which he reluctantly did.

In the meantime, Matilda invaded England on 22 September 1139. It was as if Northallerton had achieved nothing. Stephen swiftly captured her at Arundel Castle, then in an act of shear folly allowed his brother to escort her to Robert of Gloucester at Bristol. This strange act of kindness sealed Stephen's fate. She quickly rallied her supporters, including her half-brother Reginald of Cornwall, Baldwin de Redvers, Miles of Gloucester, Bishop Nigel of Ely and Brian FitzCount.

Stephen's act of chivalry was to blight the rest of his reign. He occupied himself unsuccessfully besieging Wallingford and then failed to take Trowbridge, so left a garrison in Devizes. He acted against Bristol but found it too strongly fortified and was forced to turn east to counter insurrection by the disenfranchised Bishop of Ely. The bishop hired mercenaries and fortified Ely, which was built upon an island amongst the marshes and fens. His men constructed an earthen fort, but Stephen's troops built a bridge of boats to cross the moat and routed the defenders. The bishop escaped to Gloucester where he joined Matilda.

In 1139 Stephen also besieged Leeds Castle in Kent and then campaigned in north-western England and Scotland. He seems to have taken King David's son hostage, for Prince Henry was with Stephen at the siege of Ludlow. The king actually saved the prince's life when he was nearly hauled over the walls by the defenders. Stephen captured Shrewsbury Castle and on this occasion hanged ninety-three of the rebels, including the constable. Had he maintained this policy he may have soon secured the subservience of his unruly kingdom.

In the meantime, Stephen's early treaty concessions to the Scots, especially the loss of Northumbria, continued to fuel discontent and rebellion in the north. Ranulf, Earl of Chester, seeking redress for Carlisle and Cumberland, seized Lincoln Castle. Once more Stephen trudged north at the head of an army, arriving at Lincoln on 6 January 1141 to besiege the castle. Ranulf escaped to Cheshire to raise reinforcements and Welsh mercenaries, sending a plea for help to his powerful father-in-law, one Robert of Gloucester. Both of them moved on Lincoln, arriving on 1 February 1141 with an army of at least 10,000. To get to the castle they had to cross the Fossdyke and the River Witham (the Fossdyke had been built by the Romans to link the Rivers Witham and Trent). Moving west of the city, they surprised Stephen's guard and quickly forded the Fossdyke. Once across they prepared for battle.

Stephen's supporters were divided amongst those wanting to fight and those advising garrisoning Lincoln town and retiring to collect more men. In the event, Stephen led his force, possibly fewer than the rebels, out of the west gate onto the slope leading to the Fossdyke to give battle. His army was composed of the usual three divisions, with the mounted knights and men-at-arms on the wings and the infantry in the centre. On the left were Stephen's Breton and Flemish mercenaries under William of Ypres and William of Aumale. Stephen's dismounted men-at-arms gathered round the royal standard and the Lincolnshire levies held the centre. On the right were Earls Waleran of Meulan, William de Warenne, Simon of Senlis, Gilbert of Hertford, Alan of Richmond and the unreliable Hugh Bigod, Earl of Norfolk. At Stephen's behest, Baldwin fitz Herbert, leaning upon his battleaxe, made a rousing speech claiming that Robert of Gloucester had the heart of a hare.

Ranulf and Robert's rebels were drawn up in three divisions. Earl Ranulf was with his levies and dismounted knights in the centre, Robert was on the right with the Welsh levies deployed before him, while those nobles disinherited by Stephen were gathered on the left. Two Welsh princes, Cadwaladr and Maredudd, led the levies. It was Robert and Ranulf who opened the battle.

In the face of the rebel advance, the royal right turned and ran without a fight. In fact, most of the royal cavalry deserted to the enemy. On Stephen's left, the two Williams scattered the lightly armed Welsh and attacked Earl Robert. Ranulf came to his assistance and the two Williams were put to flight. While William of Ypres and his Flemish troops abandoned

Stephen to his fate, they withdrew in good order and survived to fight another day.

Stephen's centre was left alone to face the rebels. He and his men-at-arms gave a good account of themselves, until William de Chaignes captured him. It is said that Stephen fought with such valour that he broke both his sword and a battleaxe. One chronicler recalls he battled 'like a lion, grinding his teeth and foaming at the mouth like a boar'. With a loud shout, Chaignes proclaimed, 'Here, everyone, here! I've got the King!' and Stephen's remaining supporters fled or threw down their arms. Up to 500 of them were lost in the river. The rebels celebrated by looting and pillaging hapless Lincoln.

The king was incarcerated in Bristol and to compound his misery his brother Henry of Blois sided with Matilda. King David also joined the victorious Matilda. Waleran, his Norman lands under Angevin threat, left for the continent never to return. Stephen's wife appeared before Matilda and begged in vain for her husband's liberty.

For a while it seemed the empress might become queen. In June 1141 she entered London and bribed Geoffrey de Mandeville, Castellan of the Tower, to hold Stephen's daughter-in-law. However, her haughtiness brought her to grief. She fell out with Henry of Blois and was driven from London after her punitive taxation contributed to a riot. She fled with the Earl of Gloucester to Oxford.

In the meantime, the redoubtable Queen Maud, described as a woman who 'bore herself with the valour of a man' and had a 'man's resolution', was not idle. Joined by William of Ypres, she raised an army, the city of London providing 1,000 soldiers, with which to rescue her husband. A contemporary source claimed that London could provide 20,000 mounted men-at-arms and 60,000 foot soldiers.

Even allowing for an exaggeration of a factor of 10, in the light of the contingent provided to Maud, that figure seems very wide of the mark.

Bishop Henry, having a change of heart, now hoisted Stephen's banner in Winchester and garrisoned the other castles in his diocese including Waltham and Farnham. His garrison in Winchester cleared the buildings round Wolvesey Castle and awaited the rebels. Matilda called on the Earls of Gloucester, Chester and Hereford and David of Scotland, and moved to capture Winchester Castle. She was then besieged in the city and found herself caught between Maud's army and the bishop's forces. Every building that her troops took sanctuary in was set on fire to flush them out.

Matilda, Robert of Gloucester, King David and their remaining supporters took sanctuary in the castle and were besieged for six weeks. In what became known as the Rout of Winchester, Matilda attempted to escape with Robert of Gloucester on 14 September 1141. Their pursuers overtook them at Stourbridge and Robert and his knights valiantly turned to fight. While he was captured, the empress got away to Devizes. William of Ypres took the credit for the rout and the capture of Robert of Gloucester. After this utter disaster, King David withdrew his men to Scotland.

Robert, Matilda's chief supporter and vital for her cause, was exchanged for King Stephen. The earl was then sent to Anjou to collect troops and Matilda's son, Prince Henry. On the continent between 1141 and 1145, Geoffrey of Anjou completely conquered Normandy, cutting Stephen off from his Norman allies and, preoccupied as he was, Geoffrey did not come to Matilda's aid.

Stephen was determined to revenge himself and remove Matilda's threat. He attempted to capture her at Oxford in 1142. He personally led the attack and the town fell almost immediately, but Matilda once again held out in the castle. The king brought up siege engines with which to assail the fortifications, and the defenders lasted for three months before their supplies were exhausted. Once more Matilda escaped, reaching Wallingford and the waiting army gathered by Prince Henry of Anjou and Robert of Gloucester. These forces defeated Stephen at Wilton, the king only narrowly avoiding capture for a second time.

To make matters worse, Geoffrey de Mandeville, who had been arrested in 1143, led a rebellion the following year, laying waste to the fens and East Anglia. His insurrection ended on 16 September 1144 after being mortally wounded in the head by an arrow at the siege of Burwell. Two years later, Stephen captured Farringdon Castle, which had been built by Robert of Gloucester.

These were years of stalemate. Matilda's supporters held sway in the West Country and East Anglia, while Stephen controlled London, Winchester and the Midlands. Finally, in 1148 after Prince Henry had returned to Normandy and Robert had died, Matilda gave up and returned to the continent. Even then, Stephen was not to know any peace and Matilda was to have the last laugh. Prince Henry, now Duke of Normandy, invaded in 1153 with 140 knights and 3,000 foot soldiers.

Stephen moved to face him at Wallingford but no battle took place. Shortly after Henry landed, Eustace, Stephen's oldest son, died and it was agreed the succession should now pass to Henry. After a reign blighted by terrible bloodshed, Stephen died on 25 October 1154 at the age of 50. Rather than face continued war, Henry II was crowned on 19 December. Stephen was the last of the Norman kings, for Henry was an Angevin, the first of the Plantagenet line.

In truth, Stephen's nightmare nineteen winters really only lasted six years, with anarchy occurring between 1139 and 1145. Nonetheless, he has gone down in history as a weak and inept monarch. Henry II, in contrast, controlled a realm reaching from the Pyrenees to the Solway Firth – by right of his mother, he was Duke of Normandy; by inheritance from his father he was Count of Anjou and Maine, and by his wife, he became Duke of Aquitaine. King Henry II was to recover much of the northern lands lost by Stephen, until King Malcolm IV of Scotland was only left with the Earldom of Huntingdon.

The military situation inherited by King Stephen was such that he was never ever able to fully crush his rivals, who were able to operate from the West Country and the continent with impunity. He had military superiority, but lacked the resources or grit to deliver a knockout blow and his mercenary forces were never a decisive component of his army. While Stephen's forces won a decisive victory at Northallerton, he failed to win the war against the Scots, with the result that just two years later they were back in England.

During his reign it is notable that there were just two major battles – Northallerton and Lincoln. The Conqueror's legacy of castle building meant that rebel barons were able to easily defy a weak king from their strongholds. Stephen was forced to waste resources, time and effort conducting military operations against numerous towns and castles. In some cases, he had to return to the same place several times. Even when he reduced them, his treatment of the vanquished was not a sufficient deterrent. Not until the advent of gunpowder would this situation be resolved and then it would herald the downfall of England's French possessions during the Hundred Years War.

3

Destruction of the Crusaders: Homs 1281

Despite popular perceptions, it was the Mongols, not the Arabs or Turks, who sounded the death knell of the Christian 'Crusader' kingdoms in the Holy Land or *Outremer* (the land beyond the sea). Ironically the Mongols were a far greater threat to the Near East, even after they had embraced Islam, than the Crusaders ever were.

The First Crusade of 1096 secured a foothold in the Holy Land that was to last for the next 200 years. It was the Mongols' inability to defeat the Muslim Mamluks of Egypt that eventually spelled the end of the Christian Kingdom of Jerusalem. For forty years, the Mongols tried in vain to defeat the Egyptian Mamluks. The Battle of Homs in 1281 only reaffirmed the myth that Mongol invincibility was at an end. During a hiatus in these campaigns the Mamluks conquered the last Christian stronghold of Acre, finally ending the kingdom in *Outremer*.

Two things stand out militarily during this period – siege warfare and, more significantly, the horse. The vast majority of the armies fighting during these wars were cavalry. Certainly, the bulk of the Muslim military forces were composed of horsemen. Infantry were used, but more so by the Christians (or Franks, as the West Europeans were known) defending their strategic coastal castles – although Christian knights fought mounted, their European steeds were considered inferior to the indigenous breeds.

It was the horse that helped spread Islam so quickly and gave the Middle Eastern and Central Asian warriors their significant military successes. To the Europeans, the 'Tartars' were a barbarous horde, but unfortunately for them and the Turks, the Mongol war machine was highly organised and disciplined. No armies have ever won so many battles nor conquered such huge territories as the Mongols – they fought from Poland to Indonesia.

Although Islam had controlled the Holy City of Jerusalem since AD 637, an uneasy truce existed between the two great powers in Asia Minor and the Middle East, Christian Byzantium and the Muslim Fatimid Caliphate of Egypt. Then, in 1071 a disaster befell Byzantium from which it was never to fully recover.

The Seljuks, under Sultan Alp Arslan (or 'Mountain Lion') followed by his son, controlled a huge area. It reached from Anatolia, south to Syria, the Red Sea and the Gulf, east to Khwarezmia and Afghanistan, and north to the Caucasus and the Aral Sea. The Seljuks were Byzantium's undoing.

Alp Arslan marched against Fatimid-held Damascus in preparation for conquering Fatimid Egypt. Byzantine Emperor Romanus IV seized the opportunity to assail the Seljuks' rear and regain Armenia, which had once been the empire's main recruiting ground. He gathered a massive army at Ezerum, 80 miles from Manzikert, in early 1071. It was recorded as numbering between 200,000 and 1 million men; however, most of them were little more than a poorly equipped and unwieldy rabble.

Arslan prudently withdrew from Mosul and Romanus, rather than simply restoring the old Armenian frontier, decided to attack the sultan's forces. The enormous cost of the campaign made a decisive outcome vital for the Byzantines. After much manoeuvring, the Byzantine Army of up to 100,000 confronted the sultan, with 12,000–40,000 men, on 19 August 1071 and suffered a shattering defeat. A Seljuk cavalry charge was followed by the complete disintegration of the Byzantine Army and Romanus was captured.

The Seljuks created a new state in Anatolia, the heart of the empire, called the Sultanate of Rum, or Sultanate of Roman Lands. It was clear to western Europe after Manzikert that Byzantium had forfeited its role as protector of Christendom. Such was the disaster for the empire that, from that point on, it had to rely increasingly on the dubious services of mercenaries.

Jerusalem fell to the Seljuk Turks, while war between the Fatimids (of the Shiite branch of Islam) and the Abbasid Caliphate of Baghdad (of the Sunni branch, like the Seljuks) made Syria unsafe for Christians. The scene was set for the Crusades, with the crippling dissipation of Byzantium it was only a matter of time before the West intervened. This chain of events resulted in the First Crusade and the creation of the Kingdom of Jerusalem and its neighbouring principalities.

The Near East and Middle East was to be embroiled in 200 years of warfare. Between 1096 and 1254 there were seven major Crusades, while between 1097 and 1299 there were over fifty major engagements and sieges, as well as countless smaller battles and skirmishes. The Crusaders' defeat at the Battle of Hattin, fought in 1187, is one of the most famous, however Homs was to have greater repercussions.

With the Christians now finally ensconced in the Holy Land, the Seljuks attempted to drive them from Syria. Ironically, fearing the Seljuks, the Syrians sided with the Franks. In consequence, the Seljuks were defeated at Tel-Danith in 1115 ending attempts by the Seljuk sultans of Persia to recover Syria.

However, much of the work of the Crusaders was undone on 4 July 1187 when the Franks' forces were crushed at Hattin by Saladin. This was followed by the loss of Jerusalem on 2 October, making the Frankish title of King of Jerusalem no more than a symbolic one. Two years later, the Franks only held on in Antioch, Tripoli and Tyre.

The centre of Muslim power shifted to Egypt following the death of Saladin in 1193. A Fourth Crusade was called in 1200 and after taking Zara on the Hungarian coast for the Venetians, they moved to Constantinople to help depose Byzantine Emperor Isaac Angelus in 1204. Constantinople fell, Angelus was assassinated and Baldwin IX of Flanders was elected the first Latin Emperor. The Byzantine Empire was divided amongst the Franks, with the Latin Emperor getting one-quarter and the Venetian and Frankish Crusaders the rest. The Byzantines continued to call themselves '*Rhomaioi*', or Romans, but also adopted '*Hellene*', or Greek, to distinguish themselves from the Franks.

In far-flung Mongolia in 1206, Temujin was acknowledged as the ruler over 'all who dwell in felt tents', with the title of Genghis Khan. On taking power, he had seventy of his leading enemies boiled in seventy cauldrons. From that moment on, the known world began to fear the Mongols. In 1211 they attacked China and nine years later Khwarezmia, a Muslim state covering modern Iran and beyond. A Mongol force some 30,000-strong wintered in Armenia, then crossed the Caucasus Mountains in 1221. It surprised the Georgian Army of Queen Russudan which had been gathering for the Fifth Crusade. Four years later, the Georgian capital was taken by a Muslim Army.

The Mongols returned in the mid-1230s, seeking to secure their southern flank before sweeping into eastern Europe. By 1238 Georgia had fallen into bloody disunity and the following year became a Mongol vassal. Also, the following year, the Mongols massacred the population of Ani, the ancient capital of Armenia. Between 1237 and 1242 the Mongols campaigned in Europe, culminating in a decisive victory at Liegnitz in 1241. An army of Poles, Silesians and Teutonic Knights, some 20,000 predominantly light cavalry, were crushed. Such was the carnage that the Mongols reputedly collected nine sacks of ears. The Mongols also turned on the Sultanate of Rum.

Mongol armies numbered up to 100,000, although these were increasingly multiracial. The standard unit was the *touman* of 10,000, other units included the *keshik*, or guard, and the *cherig* (auxiliaries). The *darkhat*, a class of freemen, and the *baatuts*, *noyans* and *nukuts* (hereditary noblemen) provided the officers. These forces were divided into three: the *Baraunghar* (Army of the Right Wing, or West), the *Khol* (Army of the Centre) and the *Junghar* (Army of the Left Wing, or East). The Mongol Ilkhanate of Persia had manpower resources of 100,000–300,000. Mongol tactics were very like those of the Turks, both being steppe peoples. They relied on mounted mobility, the bow, feigned flight, envelopment and relentless pursuit. Only occasionally did they fight on foot.

In 1250 the military caste known as the Mamluks overthrew the Ayyubid dynasty founded by Saladin in Egypt. The Mamluks were an almost entirely cavalry force and the Mamluk Army was organised into three elements, the Royal Mamluks, the Mamluks of the Amirs and the al-Halqa. The Royal Mamluks were essentially the sultan's guard, Mamluks belonging to former sultans and units from deceased amirs numbering 10,000–15,000. The amirs themselves were drawn from the Royal Mamluks, while the al-Halqa consisted of non-Mamluk cavalry such as Arabs, native Egyptians and Mamluks' sons born in Egypt.

The strength of these regular Mamluk troops was about 40,000–60,000. In addition, there were large numbers of auxiliaries drawn from the Turcoman, Kurd, Bedouin, Syro-Palestinian and Lebanese tribesmen. These forces could add up to another 10,000 men. Mongols were also found in their ranks after 1262, and during Baibars' reign up to 3,000 Mongols served in the Mamluk Army.

King Hayton of Little (or Lesser) Armenia hoped for a Mongol alliance against the forces of Islam. Indeed, the Armenians formed an alliance with the Mongols against the Seljuks in 1254. The Ilkhanate of Persia was established in 1256 and two years later a large Mongol Army crossed the Euphrates and defeated the Seljuks on 11 January. This army included Prince Bohemund IV, with a Christian detachment from Antioch, and King Hayton with 16,000 Armenians. This was a historic moment, for it was one of the first occasions that Christians and Mongols acted in unison. Baghdad fell on 10 February 1258 and, in a forty-day massacre, 80,000 inhabitants were needlessly butchered. The Eastern Christians welcomed the Mongols as enemies of Islam and the Armenian King saw the advance against Syria as a Crusade.

Aleppo fell to the Mongols in 1259 and suffered a mere six-day massacre. The Sultan of Syria fled to Egypt and Damascus surrendered. Inevitably, the Mongols and Mamluks were to become entangled when the Mongol Khan Hulagu demanded Sultan Qutuz's submission in 1260. This was to be a lost opportunity for the Franks, instead of allying themselves with the Mongols, in a move that may have broken Mamluk power for good, they allowed Qutuz free passage. Near Nazareth on 3 September 1260, in one of the most critical battles in history, the Mongols were defeated. The battle was fought at Ain Jalud, or Goliath's Spring, the site of David's biblical victory.

Some 12,000 Mamluks under Amir Baibars lured the Ilkhanid Army of 10,000–20,000 cavalry, including Georgian and Cilician Armenians, into a successful trap. Unreliable Egyptians fled in the face of the Mongols, drawing them into the hills where Qutuz awaited with the bulk of the Mamluk Army. At Ain Jalud, although outnumbered the Mamluks had the advantage of charging downhill, adding momentum to their attack. The Mongol General Kitbugha successfully dealt with the Mamluk advance guard and defeated their left flank.

In the meantime, the Mamluk centre and right held fast and outflanked the Mongols. Sultan Qutuz with his guard and Baibars led a decisive attack. Although they were enveloped, the Mongols fought fiercely until midday. The Mamluks allegedly pursued their broken enemy to the very banks of the Euphrates, some 300 miles away. The Mongols' reputation for invincibility was finally shattered. Kitbugha was captured, but before his execution boasted that within a year Mongolia would make good its losses in men and horses. His threat was not to prove hollow.

Shortly after the battle, Qutuz was murdered and replaced by Baibars. Having dealt with the Mongols it was inevitable that the Mamluks would turn their attentions on the Franks. Incessant internal squabbling and the demise of the Latin Empire of Constantinople after it was retaken by the Byzantines in 1261 meant the Franks' prospects were not good.

Baibars threatened Acre in 1266, while his leading Amir Qalaun invaded Cilicia. Notably the Armenians sought help from the Mongols, not the Byzantines or Franks. The Armenian King Hethum, fearing attack, went to the Ilkhan court in Tabriz. Returning with a small force of Mongols, Hethum found his kingdom and capital had been sacked. Hethum was so appalled that he eventually abdicated in favour of his son, Leo III, and retired to a monastery. Leo knew that if the price of protection was to remain a Mongol vassal, then so be it.

Slowly Baibars reduced the Frankish strongholds. In 1268 he captured Jaffa and Beaufort Castle, culminating with Antioch. In particular, the latter, with its population murdered, was never to recover. Baibars, in a fit of malice, wrote to Bohemond VI, Prince of Antioch and Count of Tripoli. He gloated:

> Oh, if only you had seen your knights trampled by our horses, your houses looted and at the mercy of everyone who passed by, your treasure weighed by the quintal, women sold in the market-place for a gold dinar. If only you had seen your churches utterly destroyed, the crucifixes torn apart, the pages of the Gospel scattered, the tombs of the patriarchs trodden under-foot.

A vital northern bulwark for *Outremer* had gone forever. Antioch had been the first Syrian city to fall to the Crusaders in 1098, and its capture after 170 years sounded the death knell of *Outremer*.

The King of Cyprus and Jerusalem was powerless and the kingdom was now confined to a handful of cities. Prince Edward, son of Henry III, arrived in Acre in 1271 with just 1,000 men. He was the first Englishman to seek aid from the Mongols. Shortly after Edward arrived, Sultan Baibars captured the vital Christian fortress of Krak des Chevaliers in Syria. This was the most formidable castle in *Outremer* and commanded the strategic valley between Homs and Tripoli.

Edward found that commerce and naked greed was thwarting any illusions of a united Frankish front. The Genoese profited from the Egyptian slave trade and the Venetians, ironically, were Baibars' principal source of weapons. Edward led a raid against the Muslims but was hamstrung without more men from Europe. In response to Edward's request, Ilkhan sent an army that captured Aleppo; Baibars advanced from Damascus but then withdrew. The sultan, in wanting to deal with the troublesome Mongols, agreed to a truce with Edward which would safeguard Acre for twelve years. However, Baibars feared Edward might successfully raise a Crusade if he was allowed to return home, so he ordered his assassination – luckily the attempt was not successful.

Once back in England, Edward continued to support the Christian alliance with the Mongols. Then, in 1277 the Mongol threat was again defeated, this time at Albistan. Baibars defeated 20,000 Mongols, Georgians and Rumi Turks with some 30,000 Mamluks, killing 7,000 Mongols and 2,000 Georgians. Baibars then withdrew in the face of a larger Mongol horde under Ilkhan Abaqa. He was subsequently poisoned and replaced by Qalaun.

During his seventeen-year reign, Bairbars undertook thirty-eight military campaigns and fought in fifteen battles. He fought nine times against the Mongols and five times against the Armenians, but most significantly his forces defeated the Franks on twenty-one occasions. Not once did *Outremer* learn its lesson.

In the meantime, the Kingdom of Jerusalem was working on destroying itself, with the strongholds of Jebail, Tripoli and Tyre at loggerheads. Acre actually celebrated when it learned that forces from Tripoli had burned Jebail after foiling an attempted coup. Sultan Qalaun found his authority contested by the Governor of Damascus, Sonqor al-Ashqar.

Mongol Ilkhan Abaqa and his vassal Leo III of Armenia called for an alliance with the Franks, but all was in vain and another opportunity was thrown away. Making the most of infighting amongst the Mamluks, Abaqa occupied Aleppo in October 1280 but his forces were insufficient to oppose Qulaun, who gathered his men at Damascus, and the Mongols retreated across the Euphrates.

A Mongol ambassador was sent to Acre to inform the Franks that Ilkhan was to again invade Syria in the spring with 100,000 men. He wanted their support, men and supplies. However, Qalaun was to pre-empt this, for he offered both Acre and Tripoli a truce, which they self-servingly accepted. The last thing the Mamluks wanted was a united Frankish front on their left flank while they prosecuted their war against the Mongols. Only the Hospitallers of Marqab were to throw their lot in with the Mongols. Qalaun also made peace with Sonqor, who had withdrawn to northern Syria.

Two Mongol armies advanced into Syria in September 1281. Once again the Franks stood by, despite the lesson of Ain Jalud twenty years earlier. They easily appreciated the consequences of a Mongol defeat and yet they did nothing. At the same time, a Mongol fleet was invading Japan, although the island was saved by the kamikaze, or 'divine wind', which destroyed their fleet.

Under Ilkhan Abaqa and his brother Mangu Timur, the two Mongol armies acted separately. Abaqa moved along the Euphrates subduing Mamluk frontier forces while Christian forces joined his brother in the form of Georgian and Cilician Armenians under Kings Dimitri and Leo III, as well as the Hospitallers from al-Marqab and Rumi Seljuks.

Qalaun should have had plenty of prior warning, for Baibars had created a system of posting houses, with relay horses, connecting the empire with Cairo. There was also a pigeon post. However, Mangu and his allies marched down the Orontes valley until they confronted Qalaun's Mamluk Army outside Homs on 30 October 1281.

Mangu's forces numbered some 80,000, although the Armenians and Georgians made up about a third. The Mongols constituted the centre and left while the allies made up the right. The Georgians, who had no love for the Franks, probably included Alans (Ossetians) from the Black Sea, who had also been defeated by the Mongols. The Armenian forces consisted of paid retainers, but probably included Turk, Persian and Frankish mercenaries. Their organisation and appearance was similar to the Franks, especially after feudalism was introduced from Antioch. The Hospitallers were initially nursing brethren for pilgrims in Jerusalem and became the Order of the Hospital of St John. By the early 1200s they were influential military brethren with garrisons at Krak des Chevaliers and al-Marqab. Their contribution was only a few knights, who joined Leo's army.

The Mamluks deployed with the Royal Mamluks, al-Halqa and Egyptians in the centre under Qalaun, and joining them was Amir Lajin with the Army of Damascus. The Bedoums and Ayyubids from Hama and Kerak on the right were commanded by al-Mansur of Hama. The reconciled Sonqor was with the northern Syrians and Turcomans on the left.

By the time of Homs, the Mamluk Army in Egypt numbered 12,000 Royal Mamluks, 5,000 Amir's Mamluks and 12,000 al-Halqa. Forces from the other vice royalties included another 3,500 Amir's Mamluks and 16,000 al-Halqa. This gave them a total manpower of over 48,000 men. Sultan Qalaun's royal bodyguard consisted of the Burdiyya Regiment, some 3,500 Circassians and Armenians. The rest of the Royal Mamluks were Kipchukis and some Mongols. Their force at Homs was up to 50,000–60,000 strong, although some sources state that up to a quarter of a million men were involved in the battle.

It is interesting to speculate whether any firearms were used at Homs, although none are recorded. The Mongols would have gained knowledge of gunpowder from China, whose first recorded recipe dates from 1044. Some form of crude artillery was used by the Mongols at the Battle of Mohi in 1241, and in their invasions of Japan in 1274 and 1281 they are recorded as using 'fire-barrels', possibly based on Chinese 'fire-lances'. By 1280, the Mamluks had an early hand cannon or naptha tubes called the *midfa*. It reportedly fired bullets or bolts, and was made of wood with a short barrel. At the siege of al-Marqab in 1285 'iron implements and flamethrowing tubes' were present.

The Mongol's Christian allies proved their worth by routing Sonqor's 12,000-strong left wing. In their bloodlust, the Armenians and Georgians chased the Turcomans and Syrians back to the Mamluk camp before the very gates of Hom. They then began to loot the camp. In the meantime, the Mongol centre successfully pressed the Mamluks. However, a Mamluk officer, on the pretence of deserting, reached Mangu wounded and unhorsed him. As this was taking place, 300 Bedoum auxiliaries wheeled round onto the Mongol left and attacked their baggage train.

Mangu appears to have lost his nerve, for he ordered a withdrawal fearing encirclement, a tactic so beloved by his own people. The Mamluks then launched a final attack. The Armenians and Georgians, on returning from their exertions found their allies gone and had to pass through the scattered Mamluk ranks. They missed a great opportunity, for the sultan only had 1,000 men around him. Qalaun, seeing this enemy body of troops, had his standards quickly lowered and drums silenced.

The Christians were to pay for their failure. Once they had passed by, Qalaun attacked their rear while the Mamluk right flank, returning from its pursuit of Mangu, also fell on them. The Armenians and Georgians paid dearly as they fought their way from the field, although Leo and some of his men managed to retreat to Armenia.

The Mongols lost most of their casualties as was traditional during the pursuit. Mamluk losses were also recorded as heavy, to the extent that they did not chase the Mongols beyond the Euphrates. Nor did Qalaun attack Armenia. The prior of the English Hospitallers, Joseph of Chauncy, was present at Homs and wrote to Edward I afterwards. He tried to deflect criticism of the King of Cyprus and Jerusalem and the Prince of Antioch and Count of Tripoli, although the former had done nothing and the latter had signed a truce with the Muslims.

With Frankish forces it is just possible that the balance could have been tipped against Qalaun. However, it mattered little, as the western Europeans were fighting amongst themselves. For the next ten years the Mamluks were able to concentrate on the destruction of the Kingdom of Jerusalem without any distraction from the Mongols.

Marqab surrendered after a month's siege in 1285, although the garrison was spared. Tripoli fell four years later, despite help from Acre and Cyprus. This time, every man was killed and the women and children taken as slaves. The city itself was destroyed.

The Mongols still sought a Western alliance which could have offered a lifeline to *Outremer*. In 1291 the Ilkhan asked England and France to send men to join a coalition that would include 20,000–30,000 Georgians. In return, if the

Mongols were to capture Jerusalem they would give it up. Nothing happened.

Only Acre, the economic and cultural jewel of the Levant, with a population of 40,000, remained a centre of viable opposition. The siege commenced on 15 April 1291 and the garrison, with fewer than 1,000 mounted men and 14,000 infantry, was outnumbered twenty to one. The Mamluks gathered outside with the armies of Egypt, Damascus and Hama: with an exaggerated 40,000 cavalry, 160,000 infantry and 200 catapults. The latter included two large weapons known as 'Victorious' and 'Furious', and smaller ones called 'Black Oxen'.

Cyprus sent 2,200 reinforcements, but all was in vain. The sultan is said to have directed 1,000 engineers against each of the city's towers, which collapsed one by one. Acre fell within a month, on 18 May, signalling the end of the Kingdom of Jerusalem. There were not enough ships for a complete evacuation and the vengeful Muslims slew every man, woman and child they could lay their hands on. The capital of *Outremer* was razed to the ground. One by one Tyre, Sidon, Beirut, Haifa and all the other fortresses surrendered to the Mamluks. Only the Christian states of Armenia and Cyprus remained as an affront to Islamic sensibilities.

The Ilkhand Mongols' conversion to Islam in 1295 did nothing to end their feud with the Mamluks. Four years later, Ilkhan Ghazan Mahmud launched another invasion of Syria. The Mamluk Army intercepted them at Wadi al-Khazindar. The Mongols were finally to get their revenge, but by then it was much too late for the Franks. The Mongol Army of up to 80,000 was taken by surprise by 20,000 Mamluk horsemen. Nonetheless, the Mongols fought a successful defensive action and drove the Mamluks off in disarray. The following day, the Mongols and their allies, which ironically included Armenians, Templars, Hospitallers, Cypriots and Georgians, finished the job, destroying three-quarters of the Mamluk Army.

In 1303 a Mongol–Armenian Army was defeated at Marj as-Saffar. However, the following year the Armenians defeated the Mamluks at Ayati. Isolated Christian Armenia was to struggle on for another 175 years.

The question remains, why did the Mongols meet their match in the Mamluks? Were the latter militarily superior? Although outwardly their forces were similar, it does appear that the Mamluks had some tactical advantages. Mamluk bows were more powerful, giving them superior firepower, and without the luxury of the Mongols' multiple mounts, due to the paucity of pasture, the Mamluks had to rely on greater horse proficiency. This stemmed from their initial selection and hard training.

Therefore, simple geography also played a role. Mamluk mounts may simply have been better suited than the hardy Mongol pony. Western Iran was perfect for supporting the vast Mongol herds and provided a base for their operations in Iraq, Syria and Anatolia. In contrast, Syria did not have horses or pasture in the abundance needed by the Mongols. Arabia was the main supplier of military mounts to Syria, which was not good horse country, and many Mongol ponies starved during the various campaigns.

Mongolian ponies were well adapted for the cold and were not really suitable for campaigning in the Middle East or southern China where the climate caused excessive sweating. This may account for the Mongols' successes in colder eastern Europe. In the Middle East, the Mongols would have employed oriental horses capable of efficient heat dissipation, but nonetheless the Mongolian pony legacy would have remained in their armies. Mongolian horses were also normally unshod, which would not have been good in Syria, although there is evidence that those of the Ilkhanate were.

Had a Western alliance been achieved with the Mongols, *Outremer*'s existence would have been extended. It is possible the Mamluks would have received a crippling blow and the Ilkhante of Persia would have remained as an ally to the Franks and the West. In failing to assist the Mongols to defeat the Mamluks, the Christian cities of the Levant sealed their own fate. Instead, the Ilkhanate of Persia passed into the Muslim camp and lasted only until 1354. In sharp contrast, the Mamluks continued to hold sway in Egypt until 1798 when Napoleon defeated them at the Battle of the Pyramids. Even so, had the Mongols triumphed, undoubtedly sooner or later they would have fallen out with the Crusaders.

4

TRIUMPH OF THE BOW:
CRÉCY 1346

In the lexicon of British battles, the names Crécy, Poitiers and Agincourt, fought during the Hundred Years War (1337–1453), stand out as triumphs of British arms. In particular, Crécy marked the supremacy of firepower over plate armour in the form of the longbow. It may also have been the first time that gunpowder was used on the battlefield in western Europe. In a single afternoon, England asserted her authority on the continent and destroyed the cream of France's knights.

However, the road to Crécy was to be tortuous, as were King Edward III's manoeuvrings to bring the elusive King Philip VI to battle. Upon the death of Charles IV in 1328, the royal house of Capet came to an end in France. He was succeeded by his 35-year-old first cousin Philip, Count of Valois, Anjou and Maine, who became Philip VI. However, Edward III of England claimed his right to the French throne was stronger, as Isabel the queen mother was the late King Charles' sister.

Edward was already a French magnate. As Duke of Guyenne (Aquitaine) and Count of Ponthieu he was one of the Twelve Peers of France. England was waiting to settle old scores with France, particularly over the expulsion from Normandy and French encroachment on Guyenne and meddling in Scotland. The duchy and the ducal capital Bordeaux were an important source of revenue for Edward. The Kings of England had held it as feudatories of the King of France since 1259.

Having only just secured the English throne, Edward was in no real position to challenge Philip. Notably Guyenne was considered a more integral part of the kingdom than Wales or Ireland. For fear of losing it and Ponthieu, he had no choice but to travel to Amiens in 1329 to pay homage to the new King of France. Philip VI inherited direct control of over only 75 per cent of his realm, with Guyenne, Flanders and Brittany remaining semi-autonomous. Under these circumstances war seemed inevitable, regardless of the dynastic claims. Philip had tasted military victory at Cassel in 1328 when he defeated a Flemish Army.

The Hundred Years War did not begin until 1337 when Philip declared Guyenne forfeit by the English. He then launched a three-year war to conquer the duchy, securing a number of towns. Preparations were also begun for an invasion of England involving 60,000 men. Meanwhile they raided from Cornwall to Kent burning and pillaging as they went. Philip sought the aid of the Genoese, who were a great maritime power, in raising a powerful fleet of 400 vessels. England sought its revenge in 1339 by torching Boulogne and thirty French vessels. Edward also invaded France from Picardy but Philip refused to give battle.

Undeterred, the following year Edward led a fleet of up to 240 ships against the French off the harbour of Sluys, in Flanders. Edward, who had long been desperate to come to grips with the French, cried, 'Ha! I have long desired to fight the French, and now I will do it, by the grace of God and St George!' Here, for perhaps the first time the Genoese cross-bow was pitted against the English longbow. Notably it was Edward's archers that gave him the advantage because of their greater range. The English roundly defeated France's 20,000-strong Grand Army of the Sea off Zeeland, ending the French threat of invasion. French losses were up to 15,000, while the English lost some 4,000. Edward laid siege to Tournai and although menaced by a French Army twice the size of his, Philip again refused to take the bait.

A bitter two-year war then followed in the Duchy of Brittany. In 1345 English forces invaded both Brittany and Gascony. The French riposte saw King Philip's son and heir, John, Duke of Normandy, with 20,000 men besiege the Earl of Derby at Aiguillon. In response to Edward's claim, Philip instigated a trade war, ordering his Flemish subjects to halt trade with England. This commerce was vital to both and resulted in Ghent, Bruges and Ypres supporting Edward. England saw Flanders as much as an ally as France courted Scotland. Furthermore, Edward found he had an ally in a banished French nobleman, Godfery de Harcourt, whose estates lay in Normandy.

Normandy had been an English possession and offered a short crossing from the Isle of Wight. Furthermore, Cherbourg and Ghent were equal distance from Paris, mean-ing the French would be distracted in equal measure. Edward was clever, for he let it be known that the army he was gath-ering in England in early 1346 was destined for Gascony. There, it was to help the Earl of Derby and relieve Aiguillon. By April some 15,000 men and 700 vessels (other sources quote 40,000: 4,000 men-at-arms; 10,000 archers, 10,000 Welsh infantry, 6,000 Irish and 1,000 vessels) were gathered at Portsmouth and Southampton ready for the invasion. Through the middle of the year contrary winds prevented a sailing. When the fleet did sail on 11 July, Edward let it be known that Normandy not Gascony was their destination. The French fleet was divided in penny packets along the French coast and offered no hindrance.

Edward landed at St Vaast la Hogue, 18 miles south-east of Cherbourg on 12 July 1346. Marching inland, the vanguard was under his son Edward, Prince of Wales, the main body under the king and the rearguard under the Earl of Northampton. It was perhaps hoped that marching in this manner would divert French forces away from their main axis of advance. An Anglo-Flemming force was also to march south-westward through Artois at the same time. On 16 July a small force of men-at-arms and 600 archers in a fleet of twenty vessels sailed to join the Flemish forces at Ghent under Count Henry of Flanders.

Edward's plan was to march on Rouen, as this was the lowest crossing point over the Seine and the quickest route to Flanders. The English Army pillaged and burned as it went. Although the Welsh and Irish mercenaries were blamed for the excesses, it was likely the English were avenging French raids on the south coast. Cherbourg, Montebourg and other towns were put to the torch.

They reached Caen on 26 July 1346 and the Bishop of Bayeux rejected a summons to surrender. Not taking any chances, he bravely withdrew into the castle with 100 men-at-arms and 200 Genoese, leaving the constable and the town to its fate. Divided by a branch of the Seine, neither the new town to the south or old town to the north were fortified. Edward's men carried the bridge of Boucherie, dividing the two halves, and the town's garrison fled. The bishop in the castle made no attempt to intervene.

The inhabitants fighting for their homes killed 500 English soldiers and the English Army went on a three-day rampage. Up to 3,000 Frenchmen were killed. Edward had wanted them all butchered, but Harcourt persuaded him not to martyr the whole town. The English navy, sailing up the Orne at Ouistreham, also captured thirty French vessels. Philip's edict of 1339 for the invasion of England was amongst the plunder. Copies were sent home and read out in every parish church.

By 25 July, Philip had gathered sufficient men outside Paris to respond, his main task being the defence of Rouen. France's military forces were now truly distracted from Guyenne and Brittany. He arrived at Rouen on 2 August as Edward was reaching Lisieux, just 40 miles away. The city's defences were secure and the bridges over the Seine removed. Edward moved down the south bank to Poissy, crossed the river and retreated north. His intention was to join his Flemish allies in Picardy but, with the bridges over the Somme also down, this proved difficult.

Edward was now in danger of being trapped, for, on the far bank of the Somme was gathered a French force supported by men from Artois and Picardy. The English made for the tidal crossing at Blanchetaque. A French force of 500 men-at-arms and 3,000 infantry, including Genoese crossbowmen, under Godemar de Fay was successfully driven off. This force has also been recorded as being over 12,000, including 1,000 men-at-arms and 5,000 foot soldiers – this seems unlikely, otherwise Edward would not have effected a crossing. Once again, the English longbow triumphed over the Genoese, who for a while successfully held a narrow pass. The main French Army, which was not far behind, captured some of the English baggage. Luckily the tide rose, halting the French, so Philip moved upriver and crossed at Abbeville.

Edward, once over the Somme, was no longer hesitant about giving battle. Besides, in the face of such a large force of French cavalry he had little choice, for it would be risky to be caught by superior French numbers on the open plains of Picardy. His route to Flanders was open if things should not go his way. He now looked for a point to turn and fight, he chose the gentle hill to the north of Crécy-en-Ponthieu. There, three terraces on the forward slope would present an impediment for the French knights.

Edward was to draw on his experience at Halidon Hill in 1333, when a combination of archers and dismounted men-at-arms defending a strong position had defeated the Scots. The longbow was largely unknown outside the British Isles; however, the French were soon to experience its devastating firepower. Edward knew that the longbow was essentially a defensive weapon, which could only play a decisive part if the enemy attacked them across the right ground.

The English Army rested in the vast forest of Crécy, having covered 335 miles in thirty-two days. On the night of 25 August, the men were given wine and meat; the officers dined with the king. The following day they moved onto the hill and mass was said, so that the English Army was physically and spiritually prepared for the engagement that was to come.

The English Army drew up with its right wing near the village and protected by the forest of Crécy; some 10 miles long and 8 miles deep, it would be difficult for the French to turn their flank. The left wing was anchored on the village of Wadicourt. The vanguard forming the right division, under Prince Edward, was drawn up towards the bottom of the slope. The Earls of Warwick and Oxford were assigned to guide the prince; Harcourt was also at his side, as were Lord Holland and Sir John Chandos. They numbered perhaps about 4,000: 1,000 men-at-arms, 2,000 bowmen and 1,000 Welsh spearmen. The Earls of Northumberland and Arundel, commanding the rearguard, formed the left division of 4,000: 1,000 men-at-arms, 3,000 archers and a number of Welsh spearmen. They were deployed further up the slope. The king took charge of the centre, acting as the reserve with about 3,000: 700 men-at-arms and another 2,000 archers.

Estimates of the English strength vary considerably from 8,000 to 14,000 – the true figure was probably about 10,000. It is also recorded by some authorities that the English Army had up to three bombards, or early cannon. Probably due to the French having such a large cavalry force, the whole of the English Army was to fight as one on foot. The horses were placed within the wagon enclosure to the rear. The archers dug holes in front of their positions to further impede the French horses and made sure they had plenty of arrows.

Utmost in the king's mind was that they should not be dislodged from their position. Edward reviewed his army, forbidding his men to leave their positions under any circumstances. He wisely placed his command post in a nearby windmill, which gave him a clear view of where the French would have to deploy. There he knighted his son, a move that greatly raised morale. At 3 p.m., with no sign of the French, the English were given leave to rest and take food and wine.

The French departed Abbeville early on 26 August, heading for Noyelles to the west of Crécy. Amadeus, Count of Savoy had reinforced Philip, with 1,000 lancers. The French sent out a reconnaissance party and four French knights discovered the English late in the afternoon. In light of the English's strong defensive position, their advice was to assemble the army, which trailed all the way back to Abbeville, allow it to rest and give battle the following day. The French finally appeared at Crécy at about 4 p.m., though it would be an hour before they made any move towards Edward. The resting English troops stood up to watch the chaos unfolding before them.

Despite the counsel of caution, the French nobles, knights and mounted men-at-arms were desperate to show their mettle. The army jostled forward out of control, shouting, 'Death to the English traitors! They will never return to England.' Philip may have finally lost his temper in frustration and authorised the attack. He could see that he outnumbered the English by a least three to one. Furthermore, Philip may have thought the English were fighting out of desperation, after all, for weeks they had withdrawn before him, crossing the Seine and Somme, giving the impression that they wished to avoid battle at all costs.

Philip's unwieldy French and allied army numbered up to 12,000 mounted knights and men-at-arms and up

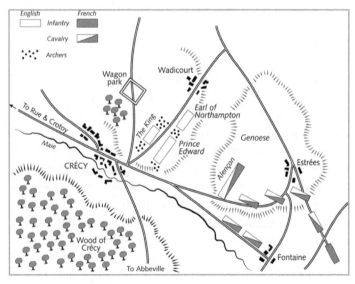

Crécy, 26 August 1346.

to 60,000 infantry, which included experienced Italian mercenaries. At a minimum, it was 30,000 strong. The mercenaries were Genoese crossbowmen numbering 2,000–6,000 and commanded by Carlo Grimaldi and Otto Doria. Crossbowmen were some of the most effective infantry in European armies by this time and those from Italian city states were much sought-after and well paid.

Riding with Philip were: Count Alençon, his brother-in-law; Jean de Luxembourg, the blind King of Bohemia, with 500 Bohemian knights and his son, Charles of Luxembourg, Emperor-elect of Germany; John, Count of Hainault, with a force of Luxembourgers; James I, King of Majorca, and the Duke of Savoy. The French may also have had several pieces of artillery, but in their rush to get to grips with the English had left them in Abbeville. The French Army has been described as a collective demonstration of armed men, with no strategy, no discipline and – vitally – no tactical plan other than the charge.

The French hoped to simply overwhelm the English, but they were poorly organised, on the whole ill-disciplined and lacked a coherent chain of command. Eventually the French and their allies found themselves in at least three divisions. First, the Genoese mercenaries with the King of Bohemia, his son Charles of Luxembourg and Count Alençon, the second under Philip, with the King of Majorca, and the third under Savoy, although there may have been a fourth under Jean de Hainault.

The first into the fray were Grimaldi and Doria's crossbowmen. It is likely that they had little say in the matter and were simply pushed forward by the momentum of the rest of the army. Certainly Philip, who had been unable to halt the pell-mell rush of his army, is recorded to have cried, 'Bring up the Genoese; begin battle, in the name of God and St Denis!'

The Genoese wanted to wait, but Alençon would hear none of it. It may be that the English scorched-earth tactics plus the treatment of the Genoese at Caen and Blanchetaque had the right psychological effect – certainly they were tired after their long march. Significantly, they were without their heavy wooden shields or *pavises*. These are mantlets with stays, which prop up giving shelter to a reloading crossbowman. The *pavises* and possibly many of their smaller bucklers (shields) and even their helmets were with the baggage train. This meant the Genoese had little or no protection from the English arrows.

The Genoese marched forward into a shower and the ground turned to mud under the feet of the entire French Army. Then, when the sun reappeared, to make matters worse it was shining in their eyes making the English barely visible. Legend has it that a flock of crows flew over the battlefield, and being a carrion bird, this was seen as a portent of evil to come.

The Genoese advanced falteringly, in some confusion against the English right and were halted three times. The English remained silent as the shouting crossbowmen plodded forward across the open fields. Giving three leaps and three loud shouts, the crossbowmen prepared to fire. Once within range the Italians let loose their bolts (or quarrels) to be met by arrows and possibly several rounds from the bombards (the first cannon to be fired in open battle).

The longbowmen did not respond until the enemy were within 150 yards, and their shafts fell in such density that they were like snow. Each archer carried a dozen arrows, with possibly as many again stuck in the ground: they could shoot off all twelve in about a minute – the crossbow could only fire, at the most, four quarrels a minute. The Genoese rate of fire would have been slow, as they would have been hampered by reloading in the mud, which would have fouled the exit channel of the bolt.

Prince Edward had some 2,000–3,000 archers. Even taking the lower figure and, say, half loosed off all their arrows, this would have been 24,000 shafts! If only half had fired just three volleys, this would have been 3,000 arrows, if they had all fired the result would have been 6,000 shafts put into the air in rapid succession – one arrow for every single Genoese. Whatever the number, the effect would have been devastating, particularly as the crossbowmen stooped to reload their weapons. It is doubtful that *pavises* would have done little more than slow down the death and mayhem. Not surprisingly under this deadly rain, the Genoese broke. Crossbows were discarded and bowstrings cut as the men fled.

Philip was furious at this setback, crying, 'Slay me these cowards, for they stop our way, without doing any good!' The Duke of Alençon was likewise furious at this turn of events, possibly fearing some deliberate treachery by the Italian mercenaries. Leading forward his knights, he rode down the hapless Genoese.

The English Army must have been highly delighted at the spectacle of allies killing each other. The Italians undoubtedly resisted and the struggling mass continued to present an irresistible target. Arrows fell thickly on the Frenchmen, Genoese and horses without discrimination, piercing armour and flesh alike. It is recorded that not one arrow missed. The English cannon would have done little damage, but the noise would have frightened the horses, adding to the terrible confusion. Ironically, the screams of their allies as they were trodden under foot deluded the French at the rear that victory was at hand and they pressed forward too.

Once clear, the knights attacked the English right, where they discovered the armour-piercing properties of the English longbow at about 60 yards. Although the French managed some hand-to-hand combat, they were repulsed. Alençon and the Count of Flanders did break through to the Prince of Wales' forces in the flank, while three squadrons of French and Germans reached the prince himself.

While this was happening, the rest of the French Army formed up and launched their men-at-arms. At about 7 p.m. the first major French charge took place across the remains of the green pasture. In reality, the charge was conducted at a trot. They had to endure the hail of English arrows but reached the English line. Under mounting pressure Harcourt, fearing for the prince's life, begged Arundel to attack and sent a messenger to the king asking for reinforcements. In light of Arundel's counter-attack, the king only sent twenty knights, for the French attack folded leaving 1,500 dead including Alençon. The king's response had been, 'Tell him from me that I know he will bear him like a man, and show himself worthy of knighthood I have so lately conferred on him. In this battle he must win his own spurs.'

There were up to fifteen poorly co-ordinated and irregular French attacks, some of which must have been conducted by the French infantry. Philip is alleged to have been unhorsed twice and possibly wounded in the face, although his actions during the rest of the battle seem to rule this out. These futile attacks lasted over two hours, and the outcome was the same each time. The English had up to 5,000 archers – 7,000, if the king's reserves were used. Three volleys on each attack would have been 21,000 arrows, and if only half the archers had fired just three arrows at each charge this would still have amounted to 10,500 shafts falling on the French. At the Burgundian Battle of Montlhéry in 1465, the ducal ledgers accounted for 38,400 arrows in one day. Between each charge the Welsh pikemen and Cornish knifemen issued forward to finish off the wounded regardless of rank or status. The order was to take no prisoners.

Most of the French attacks would have lost all impetus by the time they reached the English lines. The French squires could not reach the knights to provide fresh horses, which meant those not dead or wounded fought on foot or retired. At least one group of French horsemen, led by Jacques d'Estracelles, survived the arrows and crossed the potholes to breach the English lines on the right. They broke through towards the Prince of Wales, and Thomas of Norwich rode to the king for help.

At one stage Jean de Luxembourg, the blind King of Bohemia, being informed of how badly things were going requested to be led into battle. Surrounded by six or seven knights and several squires, he launched himself against the English. Only two squires were to survive. In the wake of the French knights' attempts, the mass of foot soldiers charged the English in an effort to finally dislodge and over-whelm them. Once more the longbow stopped the French in their tracks and the infantry staggered back under the withering fire.

John of Hainault, seizing Philip's bridle, forced him to quit the field before he was taken. The English remained secure in their position all night for fear of attack and Edward gave instructions that no pursuit or pillaging was to be conducted until daybreak. There remained a danger of being taken by surprise by contingents still coming up behind the remnants of the French Army.

The following morning was shrouded in a heavy fog. The tally of dead French knights and men-at-arms, which Edward ordered, numbered 1,541. Never had so many princes and nobles been killed – on the field lay two kings, eleven princes and eighty bannerets. They discovered the bodies of the Counts Alençon, Louis Nevers of Flanders, St Pol and Sancerre, and the Duke of Majorca and Jean of Bohemia. Estimates of total French losses vary from 4,000 to 30,000. However, the actual number was largely immaterial, what was important was that the French Army had ceased to exist. Normally most of the casualties are incurred when an army flees the field; in this case, Edward would not sanction a pursuit, so one can only assume that the vast majority succumbed to arrows.

English losses were recorded, consisting of only three knights and thirteen archers, but others put it at 300, which seems more accurate. In accordance with the custom of the day, the English claimed all the weapons, armour and much of the clothing from the dead. Once they had sorted what they wanted, so much was left over that it was piled up with some carts and burned to prevent the French retrieving it.

Philip fled not to Abbeville, where some of his army might rally, but to Amiens. Just five knights and forty-two mounted men-at-arms formed his dejected escort. He stayed there for a few days, executing some of the Genoese for alleged treachery and arranging for a four-day truce in which to bury the dead.

Northampton and Warwick were sent out with 500 men-at-arms and 2,000 archers to reconnoitre the surrounding country. They bumped into two French detachments formed by the levies from Rouen, Amiens and Beauvais who mistook the English for Frenchmen. The Archbishop of Rouen and Grand Prior of France headed the second group. It has been alleged that the English deliberately raised French standards to attract the stragglers. Before the levies knew it, the English were in their midst. They were softened up by the archers, cut down by the English men-at-arms and the survivors chased off. Nearly all the second detachment was killed, along with their two leaders. The numbers of Frenchmen dead are put at almost 8,000, but this seems improbable; although it has been claimed that the number of those killed on the Sunday were four times the number who died during the battle.

English victories were to continue. The culmination of the Crécy campaign was the capture of the port of Calais, which was to remain an English possession for 200 years. The Duke of Normandy, on hearing of Crécy, raised the siege of Aiguillon. Edward then laid siege to Calais, summoning all available cannon from the Tower of London. The city capitulated the following August. This gave Edward access to Normandy, Picardy and Artois.

In England, the Scots were defeated at Neville's Cross in October 1346, while English forces won a victory at La Roche-Derrien in June 1347, thereby securing key Breton ports and safeguarding the sea route to Bordeaux. Philip, disgraced by his defeat, died four years later in 1350, the year his navy was again defeated at the Battle of Les Espagnols sur Mer.

Throughout Europe, Crécy heralded the end of French pre-eminence. Prior to Crécy, little had been thought of English martial prowess – the French were the best in Europe. Crécy witnessed a military revolution, the triumph of firepower over armour. Indeed, the battle was a triumph for the longbow, which dominated all other weapons, both offensive and defensive. Nonetheless, the selection of the battleground was just as important for it funnelled the French onto a small frontage, ensuring English arrows could not miss man or beast. The English Army was well equipped, disciplined and spirited. Watching the farce of the French Army trying to deploy must have done wonders for their confidence. Edward's persistence had paid off. Perhaps all along, over the seven long years that he had sought to bring Philip to battle, he had known that the English longbow used in the right circumstances could humble France's finest knights.

However, French military incompetence also played a significant part. By impetuously giving battle on the site chosen by the English, the French could not bring their superior numbers to bear. They were victims of their knights' own chivalric code, which dangerously took precedence over sensible tactical planning. The French and their allies also suffered a language problem (Czech, French, German, Italian and Picard) which, combined with an ineffective chain of command, meant the French Army could not be directed in any meaningful way. The French had little military experience, John of Bohemia had fought in Lithuania, but by the time of Crécy was blind. In stark contrast, the English Army was well disciplined, motivated and led by experienced officers.

Perhaps crucially, if Philip had heeded the advice to fight on 27 August 1346 the day may have gone very differently. The thick fog would have masked the advancing French Army and made it difficult for the English archers to find their mark. In the confusion, the superior numbers of the French could have closed on the English and overwhelmed them. Instead, Philip's son John was to face defeat at the hands of the English at Poitiers.

5

WHEN ROSES CLASH:
BOSWORTH 1485

Bosworth Field has left an indelible mark on British history. Not only because it heralded the end of the Wars of the Roses (1455–85) and the succession of the House of Tudor over Plantagenet, but also because it is surrounded by unending controversy. Indeed, the great bard William Shakespeare has probably done the most to make it infamous, with his questionable characterisation of Richard III and distortion of historical events. On the military side, there has long been confusion as to the actual location of Bosworth Field and the disposition of the opposing forces, particularly that of the duplicitous Stanley brothers. The sequence of the fighting is, likewise, far from clear. The battle also marked the beginning of the pike-and-shot period, with the ever-increasing use of gunpowder and pikes on the battlefield.

The rather romantically named Wars of the Roses embroiled England in the dynastic struggles of the Houses of York and Lancaster. These conflicts had little to do with geography, and even the name stems from Shakespeare's dramatic rose-picking scene in *Henry VI, Part II*. The seeds of discord were sown upon the death of Henry V in 1422. His son, the future Henry VI, was a baby, so Henry's brothers, John, Duke of Bedford and Humphrey, Duke of Gloucester took power. Bedford was regent in France, although the Council in England refused Gloucester the full powers of Protector of the Realm. This state of affairs was to lead to constant quarrelling over the next twenty years.

When Bedford died, Gloucester was discredited after being accused of using sorcery against mentally ill Henry VI. In the meantime, Lancastrian Edmund Beaufort, Duke of Somerset, came to the fore rivalling Richard, Duke of York, who was Lieutenant of France. Beaufort replaced him and Richard was banished to be Lieutenant of Ireland, although he returned in 1450. That very year saw England lose control of Normandy after the Battle of Formigny, while the final engagement of the Hundred Years War (1337–1453) was fought and lost at Castillon three years later. This spelled the end for English-controlled Bordeaux and the Duchy of Guyenne, leaving only Calais to hold out until 1558. England's pre-eminent influence on the continent was now ended.

For the next thirty years it was English soil that was to be the battleground. There were to be fourteen major battles, ten of which were Yorkist victories. The ending of the Hundred Years War also meant that many experienced soldiers returned from France to fill the ranks of the nobles' armies. York was declared Protector of the Realm and had Beaufort imprisoned, until the king, recovering from a bout of madness, released him.

The queen, Margaret of Anjou, and Somerset called a meeting of their supporters at Leicester in 1455, then marched to meet York and his allies. This resulted in the First Battle of St Albans and the start of the Wars of the Roses. Towton, fought six years later, was allegedly the bloodiest ever fought on British soil with up to 30,000 killed (12,000 Yorkists and 20,000 Lancastrians) although more conservative and sensible estimates put the total number of dead at 7,000–8,000. This engagement established Edward IV on the throne and proved to be the most decisive of the wars.

Events leading to Bosworth Field began on 9 April 1483, when Edward IV died and England was once again plunged into the machinations of the rival noble factions. The king's brother, Richard, Duke of Gloucester acted to protect his regency and rapidly silenced all opposition. Richard had to break the Woodvilles' royal influence and there was a political power struggle to gain control of Prince Edward before his coronation.

With Henry Stafford, Duke of Buckingham, Richard seized Prince Edward at Stony Bridge and entered London with 500 men-at-arms. This was a modest force to mask the power struggle that was going on. The four main Woodville supporters, including Sir Anthony Woodville, 2nd Earl of Rivers, Edward's paternal uncle, were arrested. Richard then became Protector of the Realm. Edward IV's two sons, Edward V and his younger brother, Richard, Duke of York were declared bastards and their uncle was crowned Richard III on 6 July 1483. Afterwards, he was once again forced to silence any opposition, and the two princes were mysteriously murdered in the Tower.

Buckingham's support was rewarded with power in Wales, filling the vacuum left by the arrest of Earl Rivers. It was a move Richard was to regret. His reign was not destined to be peaceful or long. Before the year was out, Buckingham had instigated a revolt in the West Country in support of exiled Henry Tudor, Earl of Richmond and head of the House of Lancaster. However, the revolt was crushed and Buckingham executed, although Henry Tudor escaped to France.

History has been kind to Henry VII, but not so to Richard III. It is worth noting that Shakespeare did not create the physically and mentally deficient character of Richard III, but simply popularised him in his famous play. Shakespeare drew from the chronicles of his contemporaries, Edward Hall and Raphael Holinshed. Holinshed portrayed Richard as a physical and mental cripple, 'ill-featured of limbs, crook-backed, his left shoulder much higher than his right ... He was malicious ...'. Both chroniclers were influenced by Sir Thomas More's work conducted for the Bishop of Ely, a leading opponent of Richard's. Succeeding chroniclers, to gratify their Tudor sovereigns, accentuated their prejudices and Richard's deformities. Essentially Richard seems to have been a capable, honourable and just man until the time came for him to protect his position and his very life.

Henry Tudor had been an exile in Brittany since 1471 at the behest of Edward IV. In 1460 his uncle Jasper Tudor had been defeated at Mortimer's Cross and subsequently had to surrender Henry to William Herbert after the fall of Harlech Castle eight years later. The Lancastrian cause had a brief resurgence with the return of Henry VI, but the Battles of Barnet and Tewkesbury in 1471 put paid to this.

After the murder of Henry VI, Henry Tudor found himself sole heir to Lancaster. While in exile his interests were protected by his mother, Margaret Beaufort. When her second husband, Henry Stafford, younger son of the 2nd Duke of Buckingham died, she married Thomas Lord Stanley in 1482.

During the Christmas of 1482 Henry had met some 500 English exiles at Rhedon in Brittany, where he had sworn to marry Lady Elizabeth of York, daughter of the late Edward IV. This would finally unite the warring Houses of Lancaster and York. He was descended on his father's side from Owen Tudor, of the royal guard, and Catherine, the widow of Henry V. On his mother's side he was descended from Edward III through John of Gaunt, although this was an illegitimate branch.

Initially the rebellion was set for the following year and Henry sailed from St Malo on 12 October 1483 with forty ships and 5,000 men. However, he was thwarted by bad weather. His principal ally, the Duke of Buckingham, was also undone by the weather and, unable to cross the Severn, his Welsh forces dispersed and he was betrayed to the Sheriff of Shropshire. King Richard subsequently executed Buckingham in Salisbury.

Henry reached Poole with just two ships and at Plymouth news reached him that Richard was at Exeter with an army blocking his way. He withdrew and his supporters dispersed. While in Devon, Richard executed Sir Thomas St Leger amongst others and then moved to London. In a fit of pique, he also ordered all Breton vessels and their cargoes in West Country ports to be seized.

The Duke of Brittany guarded his coast awaiting possible English reprisals. Instead, a truce was reached and Richard began to negotiate for Henry's surrender. After being warned that Richard was trying to bribe the duke to hand him over, Henry fled to the French court of Charles VIII.

In light of the trauma of the Hundred Years War, France understandably still saw England as a threat and anything that kept her distracted could only be a good thing. In fact, France saw England's internal wars as God's punishment for the evil it had done on the continent. The French Council agreed to grant Henry a loan to raise 1,800 mercenaries under Philibert de Chande. Thus, Charles allowed Henry to raise a new army numbering about 3,000 English exiles and Normans. In contrast, Richard, with his coffers empty, was unable to maintain his forces and more importantly did not have a fleet with which to intercept Henry if he should try to cross the Channel again.

Lacking a standing army, Richard was obliged to rely on the nobles Norfolk, Northumberland and Stanley. Richard moved to secure Calais, for the only standing army was there and the fleet was largely designed to protect communications with the port. Furthermore, the fleet had been discredited for supporting Sir Edward Woodville. However, George Neville was patrolling the Channel and Viscount Lovell was sent to Southampton to guard the south coast. When Lord Stanley requested to return to his estates, Richard insisted that his son George, Lord Strange, must come to court, which he did on 1 August. Richard's own position looked uncertain after the death of his wife and son. Without an heir the kingdom would face more strife and this made Henry seem an attractive option.

The most significant lesson Henry and his uncle Jasper Tudor, Earl of Pembroke, learned from the failure in 1483 was the importance of making their ancestral home in Wales the centre of any future operations. The previous attempt at rebellion had simply been too dispersed, operating on three fronts. Henry's supporters, despite his setbacks, began to rally. Sir John De Vere, Earl of Oxford, who had been held in Hammes Castle in Picardy for twelve years was set free by the Yorkist governor, Sir James Blount and the pair joined Henry. This was good fortune, for Oxford was an experienced soldier and came next to Henry and Jasper in seniority. Notably Oxford had fought at Barnet in 1471. Likewise, Sir John Fortescue, Porter of Calais, and some English students from Paris University also threw their lot in with Henry.

He sailed from Harfleur on 1 August 1485 with a French and Breton fleet and some 3,000 men: a small and largely ragtag army of exiles, refugees, adventurers and mercenaries. Although the bulk of Henry's invading force was French, it also included a contingent of Scots and some Bretons. In the meantime, Richard, who was expecting trouble, established his headquarters at Nottingham Castle, as his main strength lay in the north and east. It also provided a central base from which to operate in any part of the country.

On 7 August 1485 Henry landed at Milford Haven. The success of his expedition initially hung on whether he would meet serious opposition during his progress through Wales. He avoided the stronghold of Pembroke Castle, which was held for Richard. At Haverfordwest he was joined by a handful of men led by Arnold Butler of Coedcanlas, who pledged that the men of Pembroke would serve Jasper as their earl. Henry then marched on Shrewsbury collecting supporters, predominantly from the Welsh.

It was not until they reached Cardigan that his ranks began to swell, with the Welsh gentlemen Richard Griffith and Richard Thomas joining him. Alarmingly, he also received word that a royal army under Sir Walter Herbert and Rhys ap Thomas was at Carmarthen, intent on opposing him, but this proved unfounded, although Sir Walter, an old friend of Henry's, had been sent with Thomas by Richard to stop him. In fact, the latter had promised Richard that Henry would only cross into England over his dead body. Sir Walter did not defect, but he let Henry's forces pass unhindered, and Rhys ap Thomas, being promised the Lieutenancy of south Wales, quickly changed sides.

Richard, at Nottingham, may have been taken by surprise and possibly did not hear of the landing until 11 August, some four days later. It is possible that Richard expected Henry to move through Pembroke and Carmarthen into southern England rather than through northern Wales. Regardless, the actions of Sir Walter and Thomas ended any hope of holding Henry in Wales and meant his enemy was now moving into the heart of his realm.

Rhys ap Thomas rendezvoused with Henry at Shrewsbury having followed a separate route to spare the countryside and probably keep the Welsh and French troops apart. A contemporary ballad recounts that Rhys ap Thomas had 8,000 spears with him, which is almost certainly an exaggeration. At Newport, 17 miles east of Shrewsbury, Henry's army was joined by Sir Gilbert Talbot, Sherriff of Shropshire, with 500–2,000 men, the tenantry of his nephew, the Earl of Shrewsbury, and Sir John Savage boosting their total forces to about 6,000.

Henry had a preliminary meeting with Sir William Stanley at Stafford on 17 August. Sir William was in the process of gathering his men, but Lord Stanley was at Lichfield with 2,000–3,000. However, Lord Stanley graciously withdrew down Watling Street to Atherstone in the face of Henry's advance. This effectively masked Henry's march into the Midlands.

Upon hearing this, Richard mobilised his army. His main supporters were Thomas Howard, Duke of Norfolk, with the levies of the eastern counties and his son, the Earl of Surrey; Norfolk was a veteran of the French wars and had fought at Castillon. Sir Henry Percy, Earl of Northumberland, brought his northern forces to Nottingham, Lord Lovel brought forces from London and Sir Robert Brackenbury, Constable of the Tower, arrived with his knights and some of the Tower artillery. The city of York provided 4,000 under John Hastings. Also, Francis, Viscount Lovell was called from the Cotswolds. Only Lord Stanley and his men were missing.

Lord Stanley, stepfather of Henry Tudor, bravely refused the summons, claiming he was ill in bed with sweating sickness, but Richard still held his son, Lord Strange, hostage. Therefore, Henry was granted safe passage through the Stanley lands, but they refused to support him openly. Lord Strange, after failing to escape, claimed his father was loyal, but confessed that his uncle, Sir William Stanley, Chamberlain of North Wales, had agreed to support Henry Tudor and was a traitor.

Richard marched to Leicester and then to Elmesthorpe to block Henry's march on London. At this stage, Richard may have begun to suspect Northumberland's loyalty, for he had failed to tell the city of York that the muster was to be held at Leicester. Richard must have also been concerned that twenty-eight of the realm's peers who could have joined him had not done so. Legend has it that, on leaving Leicester on 21 August, some of Richard's troops asked an old blind man by the south gate about the weather and he warned, 'if the moon changed again to-day, which has changed once in the course of nature, King Richard will lose life and crown'. Richard, passing the beggar, struck his foot on the corner post of the bridge and ominously the old man added, 'his head will strike there as he returns at night'.

Richard, secretly patrolling his army, found a guard sleeping and rewarded him with a blade in the heart, cursing, 'I find him asleep and I leave him so'. True or not, this tale probably illustrates Richard's level of agitation over the dissension amongst his so-called supporters. Certainly, he is recorded as having a bad night's sleep. Even so, King Richard was not new to battle, having been blooded at the Battle of Tewkesbury in 1471. He was wearing the same armour that had served him well on that day.

Henry, meanwhile, moved on Lichfield and then Tamworth, where his army may have obtained some artillery from the castle. In secret, on the night of 20 August, he met Lord Stanley and his brother, Sir William Stanley near Atherstone. Two deserters, Sir Walter Hungerford and Sir Thomas Bourchier, joined Henry from Brackenbury's contingent. The following day Henry was reinforced by Sir John Savage, Sir Brian Sanford, Sir Simon Digby and a considerable number of armed men. What transpired with the Stanleys is unclear, but the end result is not. Henry began to move towards Bosworth, where Richard's army was gathering.

On 21 August, Richard was camped near Sutton Cheney, about 2 miles south of Bosworth, and that evening he held a service in the church. The following day the Yorkist and Lancastrian armies prepared to do battle. It seems Richard occupied the long ridge between Sutton Cheney and Shenton. From the top, Richard could see Henry's army advancing across White Moor to the south-west, and the two separate Stanley camps to the north-west and the south-east.

The Sutton Cheney end of the ridge is about 400ft high, while the other end, Ambion Hill, is about 390ft. The crest was clear of vegetation and the slope and surrounding area was slightly marshy – it was an ideal defensive position. Richard could have drawn his army up on the flat plain near Dadlington, but the uncertain situation with his vassals made that unwise. Ambion Hill's steep northern slopes would protect Richard's right flank from Sir William Stanley, while the marsh on the gentler southern slopes would deter Lord Stanley from intervening.

The king's Yorkist forces numbered about 10,000 men (although other sources put it as high as 15,000–30,000). The actual deployment of these soldiers is rather confused. Norfolk's van, consisting of mainly billmen (the agricultural bill was used as a polearm in the fourteenth century, but from the early fifteenth century, spearmen were often termed

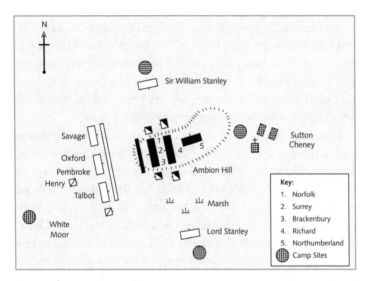

Bosworth, 22 August 1485.

billmen as well) and archers, may have numbered under 8,000, and it is thought Northumberland's rearguard numbered almost 3,000.

Richard placed a screen of 1,240 bowmen, possibly in open order to give the impression of depth, in front of the main army, with their flanks protected by two squadrons, each of 200 mounted men-at-arms or knights. Behind them were 1,000 archers, handgunners and artillerymen supported by 2,000 pikemen. The Earl of Surrey and Sir Richard Brackenbury commanded this force on the left, and the Duke of Norfolk on the right. Lastly came the main ward under Richard, 2,000 strong, flanked either side by 1,500 knights and mounted men-at-arms. The Earl of Northumberland may have commanded the cavalry on the right, but it seems more likely that he was drawn up behind Richard, supposedly to keep an eye on Sir William Stanley.

It is generally thought that Richard's army was drawn up in columns rather than lines, either because the hilltop was too narrow or because it would be easier to move against the Stanleys in the event of treachery. His forces may have included some Flemish mercenaries provided by Archduke Maximilian, who was at loggerheads with France.

It is notable that the Yorkist Army included handgunners and artillery; the latter may have been from the Tower of London. However, gunpowder was not to be an effective or decisive weapon at Bosworth Field. Artillery did not become particularly effective until the late fifteenth century, although its use in both battles and sieges goes back to the early 1300s. The guns were either constructed from iron and wood lashed together or cast from brass. Gunpowder was granulated to make it stable in the 1450s, and previously it had been mixed when needed. This meant that only cast guns could now withstand the force of the blast. Therefore, bronze barrels mounted on wheeled carriages began to appear in the late 1400s. Artillery had proved itself as a siege weapon as early as 1415 when Henry V had used ten cannon to reduce Harfleur.

France's eventual victory over the English was attributed to her embracing gunpowder and artillery. Certainly, the English longbow met its match at Formigny in 1450, when two French cannon were able to bombard the English archers from beyond bowshot. French artillery had also assured the reduction of England's strongholds in France. The 250 French siege guns at Castillon had contributed to the defeat of the English relief force. However, proper field artillery probably did not play an effective part on the battlefield until the Battle of Fornova in 1495. Handguns, in effect small cannon fixed to a wooden stock, came into general use in the 1380s and were fired bazooka fashion. By the mid-fifteenth century, the lighter arquebus had appeared which could be held against the shoulder. They were only effective in volleys at close range and the rate of fire was very slow.

In theory, the Yorkist forces included a further 4,500–6,000 men. Sir William Stanley was deployed to the north of the Yorkist Army with 2,500 men (some sources quote 3,500), while Lord Stanley may have been deployed to the south with a further 2,000 mounted infantry (other sources 2,500), but this is uncertain. Richard expected Sir William who, unlike his brother, was thought to be loyal, to join him.

Henry's Lancastrian Army numbered up to 8,000, including 2,000 Frenchmen, many of whom were probably handgunners. A skirmish line of English, Norman and Welsh archers was deployed in front of the army. The Earl of Oxford commanded the centre, while Sir Gilbert Talbot commanded the right and Sir John Savage the left, where Sir William Stanley was expected to join them. Henry and Jasper were in the centre. Jasper had previously fought at St Albans, Ludford Bridge and Mortimer's Cross. The Earl of Oxford, in an effort to retain his formation, ordered his men not to advance any further than 10ft in front of his Silver Star standard, a lesson he had learned at the Battle of Barnet.

Despite his numerical superiority Richard knew his army faced widespread disaffection and the Duke of Norfolk was his loyalist ally. From the disposition of Henry's army, he clearly expected the Stanleys to join his centre, and their reticence had severely weakened it, leaving Oxford outnumbered.

Thus, on the day of Bosworth Field there were in reality three armies. Richard's central force of about 10,000 only moderately outnumbered Henry. Nevertheless, Northumberland's commitment was suspect, which potentially meant the two were equally matched. If anything, Henry's mercenaries may have offered him a slight edge. The third army was that of the Stanleys with their 5,000 men. If they were indeed divided either side of Richard's army they were in a weak position for treachery, for either force could be defeated piecemeal by their larger neighbour.

In a way, Henry's situation was almost that of Wellington's at Waterloo where the Allies awaited the Prussians. Henry had to potentially win the day on his own or hold out until such time as the Stanleys felt it prudent to come to his aid. In the back of his mind, he must have feared that the Stanleys would side with Richard if things went against him. Lord Stanley may have provided Henry with a token force of cavalry while his main force remained aloof, despite Henry's appeal that his vanguard join the fray immediately.

Richard was eager to attack and overwhelm Henry's smaller forces, but this was contingent on rapid support from the Stanleys. He had sent a message to Lord Stanley ordering him to act without delay and bring up his soldiers by dawn, or his son's life would be forfeit. Stanley's reply was a curt refusal and a reminder that he had other sons. In a rage, Richard ordered Lord Strange to be executed but his guards refused to comply. His execution may have been deferred until his father and uncle were in Richard's custody.

The initial exchange with crude artillery, handguns and arrows achieved very little. Oxford then advanced scattering the Yorkist archers. Norfolk counter-charged with his men, and his son, the Earl of Surrey, on the left, also advanced with a wedge-shaped phalanx. It was not until Henry's forces had passed the marsh that Richard ordered his attack. In this respect Norfolk lost an opportunity, for he should have attacked while Henry's army was forming up, particularly as it wheeled northwards past the marsh.

The missile exchange was followed by hand-to-hand fighting. Suspicious of some ruse by Oxford and his tight ranks,

Richard's men hesitated. Seeing this, Oxford, with Talbot and Savage supporting his wings, attacked in wedge formation. It may have been at this point that Norfolk was killed and his son captured.

The clash of arms in the centre was to last about two hours. In the midst of the brutal melee, Norfolk and Oxford came face to face. Norfolk's sword hit Oxford on the hand, but the return blow caught Norfolk's helmet, knocking off the visor. The unfortunate knight was then promptly struck in the face by an arrow. Surrey took command of the Yorkist centre, and Oxford's infantry were forced back. Henry, meanwhile, was well to the rear of his army and was to remain so.

It seems, however, that the Yorkist centre began to be outflanked. Sir Gilbert Talbot, on the Lancastrian right, led his 2,000 Shropshire levies forward in an attempt to seize or kill Surrey. Sir William Conyers and Sir Richard Clarendon moved to oppose him, only to be killed by some of Sir John Savage's men pressing in from the Lancastrian left. Surrey, despatching a lowly levy, was finally forced to yield to Talbot.

In the meantime, Richard, Northumberland and the Stanleys had all remained inactive. Northumberland stayed put, possibly on the pretence of guarding against an attack by Lord Stanley. The following sequence of events are rather confused, and it is not exactly clear whether the Stanleys attacked before or after Richard had moved, although it does seem most likely after. Richard spotted Henry's banner moving towards Sir William Stanley, so he decided to travel behind his army and attack Henry, as success would surely finally sway the Stanleys to his cause.

With about eighty knights, Richard led a desperate charge against Henry Tudor's escort. His attack was launched on his right, past Sir William. It appears that it was about now that the Stanleys began to advance, certainly Sir William did so. In the struggle that followed, Richard proved his prowess as a knight. He made three desperate cavalry charges, and at the third, as Norfolk was cut down, he reached Henry's bodyguard. Spurring his horse into the Lancastrian Army and towards Henry, Richard succeeded in lancing the royal standard-bearer, Sir William Brandon, who fell with the Red Dragon of Cadwaladr banner. Sir Richard Percival, despite a ghastly leg wound, managed to retrieve the banner. Meanwhile Richard, wielding his battleaxe, hewed down Sir John Cheney and Sir Gervaise Clifton. Whether Richard and Henry ever came into contact we shall never know.

By now, Sir William Stanley's knights had crashed into Richard's force cutting him off. Sir William Catesby pleaded with Richard to flee, but it was too late. The king's charger stumbled in some marshy ground and was killed, leaving Richard stranded on foot, no doubt crying something similar to Shakespeare's line, 'A horse! A horse! My kingdom for a Horse!' In the confused fighting, the king was cut down either by the Lancastrians or Sir William's men who had betrayed him. He may even have been hacked to death by Henry's Welsh pikemen, his horse stuck fast in the marsh. Regardless, he fell dead upon the mud, his body mutilated. Richard was the first King of England to be killed in battle since 1066 and was the last of the Plantagenet kings who had ruled since 1154.

Stanley's force was now fully committed, and his archers began to pelt many of the Yorkists. Also, Lord Stanley, to the south, possibly committed his men and so the Yorkist Army came under pressure from three sides and collapsed. Lord Strange was abandoned by his guards towards the end of the battle, escaped and joined his father. Brackenbury remained to fight it out with Sir Walter Hungerford, who had deserted to the Lancastrians before the battle. Northumberland's troops, who had not struck a single blow, after witnessing the Stanleys' action, retired north. In fact, it is thought only half of the men on the field were committed to battle.

Upon Richard's death all his men threw down their arms and surrendered. Many are alleged not have had the stomach for the fight in the first place. The engagement had lasted about two hours, although the pursuit continued to Stoke Golding. The total numbers killed vary from 1,000 to 4,000. Conservative estimates claim the Yorkists suffered about 900 casualties and the Lancastrians 200–300. Lord Stanley discovered Richard's crown lying under a bush and placed it on the head of the victorious Henry Tudor, soon to be Henry VII upon his coronation on 30 October 1485.

Richard's corpse, stripped of its armour, was unceremoniously flung across a horse and taken to Leicester, his head striking the post on the bridge as it went. The nuns of Grey Friars, of whom Richard had been a patron, pleaded for the body. It lay exposed for two days as proof of his death before they were allowed to bury it in their church. The building was eventually destroyed and it was thought that Richard's bones were dug up and his coffin repurposed as a horse trough. However, this was found to be untrue when his remains were located under a car park in 2012 in the same area.

Henry VII's early years were fraught and Bosworth did not really end the politicking of the Wars of the Roses. Like Henry IV, Henry VI, Edward IV and Richard III, Henry Tudor was to discover that to hold onto his throne would require the continued services of an army. Within a year of Bosworth there was a short-lived rising in the North Riding of Yorkshire led by Lord Lovell (Richard's chamberlain and admiral), Sir Thomas Stafford and his brother Sir Humphrey (who had been disinherited in the wake of Bosworth and hidden in Colchester Abbey). However, their forces dispersed at the approach of Henry and Lovell fled to Flanders. He eventually sailed to Ireland to support Lambert Simnel, who was impersonating the Earl of Warwick (he was actually in the Tower of London). Simnel was crowned Edward VI in Dublin in May 1487 and he and his supporters invaded England the following month.

Henry marched north with a force of Bosworth veterans including Jasper, Rhys ap Thomas, Cheney, Hastings and Brandon. Northumberland, now reinstated, gathered 4,000 reinforcements for the king, but was forced to return to York after Simnel's supporters menaced the city: he was to be murdered in 1489. The rebel army of some 10,000 men, including 2,000 mercenaries, drew up to the south-west of East Stoke. Oxford, with the 6,000-strong vanguard coming upon them unprepared, attacked immediately. Jasper arrived just in time to prevent him being overwhelmed and the rebels were destroyed. Lovell disappeared and Simnel was captured, becoming a faithful servant in Henry VII's household.

Despite the decisive nature of Bosworth, the Wars of the Roses did not end until Henry had defended his crown from Yorkist claimants at Stoke in 1487 and, to a lesser extent, Blackheath in 1496, but military and politically a new period was beginning in British history. Indeed, Stoke was the last pitched battle associated with the Wars of the Roses.

The execution of Edward, Earl of Warwick, (heir to the House of Plantagenet) and Perkin Warbeck, the following year, truly marked the end of the Wars of the Roses. The year 1485 was, in effect, a watershed and the House of Tudor would rule for the next 120 years. Gunpowder and firearms would increasingly make their presence felt on the battlefield.

6

Rape of the New World: Tenochtitlan 1521

The tragedy of Tenochtitlan makes the fall of Constantinople almost a hundred years earlier look a much lesser calamity. How did a tiny army of several hundred Spanish conquistadors with 50,000 lightly armed Indian allies bring the very heart of the mighty Aztec Empire to its knees so quickly? Within the space of a year, what had taken 200 years to build was rubble, expunged from the face of the earth.

The Spanish certainly had the advantage of steel, gunpowder and the warhorse, but they also inadvertently waged debilitating biological warfare on the superstitious Aztecs. Rarely has the impact of a disease been so militarily decisive.

The Aztec city fortress of Tenochtitlan, now Mexico City, was founded on an island in the southern end of Lake Texcoco (actually five interconnected lakes, with Zumpango and Xaltocan to the north and Xochimilco and Chalco to the south) in about 1325. It was established by Chief Tenochtli, or Stone Prickly Pear (*tetl* – stone, *nochtli* – prickly pear). His tribe, called 'Mexica' (hence, the Valley of Mexico), are believed to have come from the Aztlán region (hence *Azteca* – person of Aztlán). He ruled for fifty-one years until his death in 1375 when the first Tlatoani, or Great Speaker, was appointed.

The Aztec Triple Alliance (the cities of Tenochtitlan, Tlacopan or Tacuba on the western shores and Texcoco on the eastern shores of the lake) conquered all their neighbours until civil war broke out in the 1460s. The Aztecs had secured their dominance by 1473, with Tenochtitlan absorbing the northern adjacent city of Tlatelolco, which had been built on the same island. The principal centres of the Aztec capital became the great temple and square of Tenochtitlan and the great temple and square of Tlatelolco. The latter was to be the final point of Aztec resistance to the Spanish siege in 1521.

Each successive Aztec ruler vanquished new vassals to provided tribute to Tenochtitlan. Between 1375 and 1502 the Aztecs recorded in the *Codex Mendoza* conquering at least 145 towns and cities. Tribute included large quantities of cotton, grain, maize and other foodstuffs, as well as military apparel (wardresses, shields and weapons). Clothing tribute was particularly important, as cotton could not be grown in the area because of the high altitude. The lords of Tlatelolco were obliged to pay tribute to Tenochtitlan every eighty days, except wardresses, which were supplied once a year. Those provinces who paid tribute on time were left largely autonomous; those who had resisted the expansion of the Triple Alliance had to supply a much greater tribute.

Some 200 years after its establishment, Tenochtitlan was a sprawling metropolis 5 miles square, with up to 150,000 souls. This was easily twice the population of contemporary London or Paris. About 7,000ft above sea level, the city was joined to the mainland by three main causeways, in fact Greater Tenochtitlan had expanded to the shores of the lake. Two aqueducts supplied the city's fresh water, one of which was 3 miles long. Dubbed the Venice of the New World, its streets were indeed formed by many canals. Tenochtitlan's Spanish conqueror declared it, 'The most beautiful city in the world'.

Montezuma (Brave Lord) II was appointed ninth Tlatoani in 1502. He found himself ruling an empire that covered an area larger than France and included most of Mexico's 11 million people. The Aztec heartland may have numbered up to 1.5 million. Montezuma was, in fact, an elected ruler and not a hereditary one, being chosen by the Aztec High Council. Furthermore, in reality his empire was really a very loose confederation of bickering vassal city states and this was to be the Aztecs' undoing.

Created by war, the Aztec state was continually at war as the component parts sought to break away. It was capable of raising armies of between 20,000–200,000, although it had no standing forces. The peasants were called up on a rota system, while a royal bodyguard and the Knights of the Eagle, Jaguar and Arrow formed a cadre of regulars. The Aztecs had no steel weapons; they had copper, but no alloys such as bronze. Their arms were wooden with obsidian cutting edges; they also used bows and a spear thrower to propel darts and javelins.

The Aztec way of war was highly ritualised and fought in a manner totally alien to Europeans, who knew that to defeat an enemy you must destroy him. The Aztecs wanted tributary vassals. Even when not at war, the Aztec conducted 'flowery wars' designed to capture sacrificial victims. Their way of life thrived on slavery and human sacrifice. The Aztec universe had to be perpetuated and this required constant blood.

The eighth Tlatoani (1486–1502) inaugurated the great pyramid of Tenochtitlan in 1487 with the appalling sacrifice of 20,000 prisoners at fourteen temples over four days. These sacrifices were designed to appease their principal deity

Huitzilopochtli (the Hummingbird Wizard), god of sun and warfare. Another god, Quetzalcoatl (the Plumed Serpent), god of learning and the priesthood was believed by the Aztecs to have been driven east over the sea by Huitzilopochtli. He promised to return in the year *Ce Acatl* (One Reed) to reclaim his kingdom. Fatefully, *Ce Acatl* fell in 1363, 1467 and 1519. In the years preceding the Spanish invasion, Montezuma argued with his old friend the King of Texcoco, who predicted doom and gloom for the year 1519. His death in 1516 was a blow to Montezuma, for while it strengthened Tenochtitlan's position in the Valley of Mexico, it ultimately and fatally weakened the Triple Alliance.

Montezuma was a brave warrior, distinguishing himself in his youth and gaining the highest military rank. During his reign he conquered over forty major communities. Despite this, a major failing of his was dealing with the city-state of Tlaxcala and its two major allies, the Cholulans and the Huexotzinco. In 1507–08 a feud broke out between the Huexotzinco and the Tlaxcalans, the former were defeated and sided with Tenochtitlan for protection. The Cholulans also entered into a truce with Tenochtitlan, an act that they were to deeply regret.

Even so, the combined forces of the Triple Alliance were unable to subdue the Tlaxcalans and by 1518 Tlaxcala had restored relations with the Huexotzinco. This failure to crush Tlaxcala was to have major implications for the Triple Alliance and Tenochtitlan. By the time of the Spanish invasion, Tlaxcala was a formidable military power with a formidable grudge against its wealthy neighbour. This and Montezuma's vacillation and lack of initiative in dealing with the Spanish were to cost him dearly.

In the late 1400s and early 1500s a proficient military had been stimulated in Spain by the Reconquista, which drove the Muslim North African Moors from Iberia, and the Italian Wars (1494–1559) against France. This, combined with the impetus created by Columbus in the Caribbean, meant that Spain was on a collision course with the pre-Columbian civilisations in Central and South America.

Between 1499 and 1508 Spanish expeditions had been exploring the northern coast of South America. In the Caribbean, Hispaniola (Haiti) was secured as a base of operations against Cuba and the Antilles. Hernán Cortés arrived in Santo Domingo, on Hispaniola, in about 1504. Initially he had intended to fight in the Italian Wars like his father, but had drifted to the New World. Five years later, he joined Diego Velázquez in conquering Cuba. For his success Velázquez was rewarded with the governorship of the island and Cortés was appointed his secretary.

Then, in 1517, Francisco Hernandez de Cordoba landed on the Mexican Yucatán Peninsula. He received a mixed reception from the declining Mayan civilisation, which ended in twenty dead Spaniards. For the time being, the Maya were to be spared the rapacious Spanish. The following year an expedition sailed into the Gulf of Mexico, landing in what is now Tabasco. Juan de Grijalva, Velázquez's nephew, made contact with the Totonac people who told him of the wonders of Tenochtitlan and, importantly, how they resented Aztec authority.

Meanwhile in Tenochtitlan, Montezuma's spies reported on the progress of the strange visitors from across the sea. It seemed the legend of Quetzalcoatl was coming true and he hoped that he could conclude his reign before the transition of power. The emperor resolved to meet these visitors with the utmost hospitality. First gold and precious stones were sent to the Spanish, who returned with them to Cuba. Their greedy appetites wetted, they were not slow in returning.

Cortés set sail from Cuba with 550 soldiers, sixteen horses, ten brass cannon and four smaller ones on 18 February 1519. Most of the men were hardy Castilians from the dry tableland of central Spain. His lieutenant was one Pedro de Alvarado, who was from the same province as Cortés. However, Cortés nearly never left. At the last minute, Governor Velázquez, having appointed Cortés captain general of the expedition but fearing the success of Cortés' preparations, tried to remove him; but it was too late, he had already sailed. Despite claims that this expedition in the New World was for the glory of God and Spain, the true motive was greed – greed for gold.

Cortés headed for Tabasco and once ashore his muskets and horses quickly defeated a Tabascan army of up to 12,000. Thus, at Potonchán on the mouth of the Tabasco River, Cortés experienced his first major victory when his European arms drove off the hostile Tabascans. His cavalry attacked the Indians from behind, taking them by surprise. The natives had never seen horses before, to them they were some sort of supernatural beast and many fled in abject terror.

The Spanish then sailed further up the coast, landing on 21 April 1519, not only the same year and same month as the Aztec prophecy, but the very day. Once on the mainland, Cortés founded Vera Cruz and a municipality answering direct to Spain and snubbing Velázquez's Cuban authority. In a stroke of luck Cortés discovered a shipwrecked Spaniard who spoke Mayan, and an Indian woman, part of the spoils from the Battle of Tabasco, who could speak Aztec: the latter, Malinche, quickly learned Spanish and became his mistress.

Once again, Montezuma's gifts, which were intended to persuade Quetzalcoatl to leave so that the Tlatoani could complete his reign, had the reverse effect. Cortés, defiantly sinking all but one of his ships, marched on Tenochtitlan. His army of some 400 men, fifteen horses and six cannon faced a journey of 200 miles. His forces also included some 3,000 Carib, Mayan, Tabascan and Totonac Indians. The Totonacs advised him to march via Tlaxcala and provided forty warriors and 200 porters. The forthcoming Tlaxcalan campaign, after the rehearsals with the Tabascans, was a taste of things to come and in hindsight proved crucial to Cortés' overall success against the powerful Aztecs.

En route, the Tlaxcalan Indians lured him into a trap involving their army of 30,000–100,000 (although a truer figure is probably nearer 15,000). It is likely that the Spaniards were victims of local politics, for the Tlaxcalans hated the Aztecs and their vassals. The Tlaxcalans sprang their trap on 1 September 1519 involving 3,000 warriors, but they were driven off. Undeterred, they launched a second attack the following day involving two armies about 6,000 strong.

The Battle of Techuacingo lasted an hour despite the Spaniards' muskets and cannon. Cortés and his men were driven back and besieged on a hilltop for two terrifying weeks. He and his men issued forth on 5 September and were attacked by up to 50,000 Tlaxcalans, so they withdrew to their camp. At one point, 10,000 warriors launched a night attack, but once the alarm was sounded Cortés sallied out and drove them off.

Unable to defeat the invaders' modern weapons, the Tlaxcalans finally sued for peace and became very firm allies. In another stroke of luck, Cortés found himself again befriended by a people who had never been conquered by the Aztecs.

On their approach, Montezuma invited the Spaniards to Tenochtitlan via the sacred city of Cholula. For the Aztecs, Cholula was Tenochtitlan's first true line of defence. Fearing that it was a trap, Cortés took 6,000 Tlaxcalan warriors with him. In turn, such an alliance can only have fuelled Aztec speculation over the Spaniards' true motives.

Worried that he and his men could not tell the warring Indian tribes apart, Cortés got the Tlaxcalans to wear a recognisable field sign. They opted for plaited garlands of feather grass. Diplomatically, Cortés instructed the Tlaxcalans to camp outside Cholula. Once in the city, he requested the Cholulans provide 2,000 men for his advance on Tenochtitlan. However, word soon reached him that the Cholulans allegedly intended to trap the Spanish between their gathering force and Aztec troops sent by Montezuma. He was warned that Montezuma had sent an army of 20,000 to Cholula, half of whom were now already in the city.

Before they could act, Cortés and his allies dealt ruthlessly with the Cholulans and it was now the Cholulans who were trapped by the Spaniards as the Tlaxcalans fought their way in to join their newfound friends. Cortés' allies went on the rampage for two days and some 3,000 innocent inhabitants were killed.

The Tlaxcalans, instigators of the massacre, were probably settling old scores with their neighbours. In light of his subsequent actions at Tenochtitlan, one can only speculate whether Alvarado was also partly responsible. He was described as reckless and volatile, certainly not a good combination in the face of the stressful and delicate situation facing the outnumbered Spaniards.

Encouraged and emboldened by Cortés' success, the Tlaxcalans now offered an army of 10,000 warriors for the advance on Tenochtitlan. Foolishly he turned them down, taking just 1,000 porters. At this stage Cortés may have been hoping to co-opt the Aztecs as he had done with the Totonacs and Tlaxcalans. Taking a large force of their traditional enemies would not have helped his cause; also, at this stage he would have not known just how trustworthy the Tlaxcalans were.

The conquistadors crossed the mountains on 2 November 1519 to be confronted by thirty cities spread out before them. It had taken them five months to reach the Valley of Mexico. Cortés had arrived at Tenochtitlan at the end of summer when the harvest was foremost in the Aztecs' minds rather than war and Montezuma greeted them warmly six days later:

> You have come back to sit on your throne, to sit under its canopy. The kings who have gone before, your representatives guarded it and preserved it for your coming ... This was foretold to us; you have come down from the sky. Rest now, and take possession of your royal house. Welcome to your lands, my lords!

Even after the glories of Madrid, Constantinople and Rome, the Spaniards were in awe of Tenochtitlan, but inside the city the daily human sacrificial rituals that took place on the summit of the great pyramid soon horrified the Catholic Spaniards. In the ceremony the heart was cut from the victim, the body tossed from the temple; the limbs went to the priests, the torso to the neighbouring zoo. The heads were displayed on racks (*tzompantli*) until they rotted and fell off. One such rack was said to contain 136,000 human skulls. The whole aspect of Aztec human sacrifice can only have convinced the Spanish that this was the work of the Devil designed to taint the paradise of Tenochtitlan. They were used to the excesses of the Inquisition, which used barbarity to unmask heretics, but this was something else.

Vastly outnumbered in the midst of this strange city, the Spanish cannot have felt safe. Across the lake the city of Texcoco was not happy at their presence and its leaders were captured as a precaution. Nor was it long before Cortés seized Montezuma himself. The emperor's arrest was based on his alleged complicity

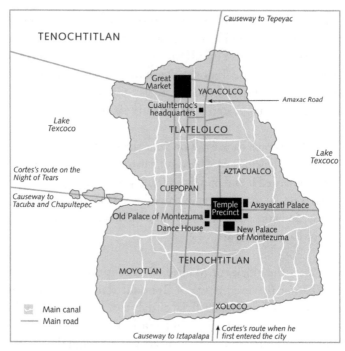

The city of Tenochtitlan, 1521.

in the murder of two Spaniards at Cholula. Despite having a bodyguard of 200 nobles, when faced with death Montezuma surrendered himself without a struggle and was moved to the palace occupied by the Spanish. Cortés made him swear allegiance to the King of Spain and treated him as a puppet.

The façade of normality did not last. Rebellion began to ferment amongst the distrustful Aztecs and, to make matters worse, Velázquez had sent soldiers from Cuba to reassert his control. This force was quite sizeable, numbering 900 men including eighty cavalry, eighty arquebusiers and fifty crossbowmen. They were commanded by Pánfilo de Narváez, a personal friend of Velázquez. He had served in the campaigns in Jamaica and Cuba; in particular, he had gained a bloody reputation during the conquest of Cuba, allegedly killing 2,000 Indians himself. It was clear that Velázquez meant business and that Cortés' impertinence was not to go unpunished. If this small army were left unimpeded it would cut the conquistadors' lines of communication to the coast. Cortés had to decide, Tenochtitlan or Narváez?

Cortés, leaving Alvarado in charge in Tenochtitlan, marched with just eighty-two men on Cempoala to deal with the Cuban force. Collecting men from Cholula, Tlaxcala and Vera Cruz, his small army grew to 265 men. The Tlaxcalans also provided 600 warriors, but most quickly deserted at the prospect of fighting Spaniards and Cortés released the rest. Despite its numerical superiority, Narváez's force was quickly defeated amid allegations of treachery on 27 May 1520. The surviving Spaniards were incorporated into Cortés' army as much-needed reinforcements.

In Cortés' absence Alvarado, with 150 men, found himself besieged in the city and under violent attack. He had brought this situation upon himself in trying to tighten his hold on Tenochtitlan. First, he had arrested the military leader of Tlatelolco and killed several other important chiefs, then during the feast of Toxcatl Alvarado's soldiers had massacred 600 of the Aztec worshippers before taking refuge in the royal palace. Perhaps Alvarado was just following orders and the tactics of terror used at Cholula. Whatever the reasoning, the population of Tenochtitlan rose up against the invaders. A desperate plea for help was sent to the coast.

Strangely, Cortés' return journey was uncontested and he re-entered the city on 24 June 1520 with 1,250 Spaniards and this time supported by an army of 8,000 Tlaxcalans. Alvarado and his comrades' relief must have been palpable. The lack of opposition showed the Aztecs' complete inability to grasp the most rudimentary European tactics or strategies. The Aztecs could simply not countenance the piecemeal destruction of an enemy's army. It also revealed the political chaos amongst Tenochtitlan's ruling elite.

The lull in the fighting was short lived. The Aztec High Council, having selected Cuitláhuac, Montezuma's brother, as successor, ordered an immediate attack. The Spaniards and their Tlaxcalan allies were trapped in the Palace of Axayacatl. Cortés' gamble had not paid off, for his army was simply too small to subdue the city's vast population. Montezuma's authority had evaporated. The fury of the assault surprised even the Spaniards, whose muskets, cannon, crossbows and swords cut down up to forty warriors in each charge.

On the third day of hostilities, Cortés made the captive Montezuma address his people from the palace roof. The Spaniards wanted an end to hostilities and safe conduct from the city. Montezuma's appearance initially brought the fighting to a halt, but was quickly met by a volley of missiles, which allegedly killed the former emperor. It is rumoured that Cortés had him murdered once he had out-lived his usefulness.

Only six days after his return, Cortés decided to abandon the city and fight his way out, heading west to Tacuba. The causeway, however, had three of its bridges missing so a port-able replacement was built. Before the break-out, Cortés offered his men the Aztec treasure saying, 'Take what you will of it. But beware not to overload yourself.' Many of Narváez's former men did not heed his advice.

Just before midnight on 30 June 1520, on what was to be known as La Noche Triste (or the 'sad night'), the Spaniards and their Tlaxcalan allies crept from the palace. Soon, great booming war drums brought both the city and lake alive with furious Aztec warriors.

In the fighting that followed the portable bridge was abandoned, and many Spaniards burdened with gold were drowned or dragged off kicking and screaming for sacrifice. Cortés reached the mainland and then discovered that Alvarado and the rearguard were in trouble so he galloped back to rescue him. The pair fought their way to the far shore, but two-thirds of Cortés' army did not make it. Tragically 270 men never received the order to move and were left trapped within the city. Over 600 conquistadors were slain along with several thousand Tlaxcalans in this calamitous retreat. Most of Montezuma's looted treasure ended up in the lake. Order was not restored until they reached Tacuba. Malinche survived the ordeal and – importantly – Cortés' shipbuilder, Martin López, who had built four brigantines to survey the lake.

The Aztecs once more threw away a chance to defeat the Spaniards once and for all. Instead of giving chase, they contented themselves with reclaiming their lost treasure and sacrificing the hapless prisoners. The survivors had to cross 100 miles of Aztec territory and for a week their retreat was unimpeded. Then, on 8 July 1520, at Otumba, the Texcocans set upon the Spaniards. The surrounding valley looked as if it was covered in snow as the sun reflected off the Indians' cotton armour. It is claimed the Texcocan Army numbered 200,000, although the number is much more likely to have been nearer 20,000.

The disciplined Spaniards and their allies retained their ranks, allowing only a small number of Indians to engage them at any one time. Moving forward, however, things soon began to look bleak and the Spaniards were only just saved by their small force of twenty-three cavalry. After a fierce battle, Cortés managed to kill the Indians' leader, cacique Cihuaca, and the Texcocans fled. This was a remarkable feat for the tired and hungry Spanish-Tlaxcalan forces. On this particular occasion the horse and lance had proved the decisive weapon.

Once back in Tlaxcala, a furious Cortés waged two campaigns to pacify the eastern coastal area as well as securing the local Spanish forces. First against the town of Tepeaca for killing Spaniards; then with 313 Spanish troops and 30,000 Tlaxcalans, he moved against an Aztec blocking force also 30,000 strong at Mexinca. Cortés fought a battle at Huaquechula against 3,000 Montezuma supporters and afterwards drove another 6,000 from Izúcar. Then, he and his comrades plotted their revenge, although this had actually started without them. In the Spaniards' absence the Valley of Mexico and Tenochtitlan was stricken by smallpox, a gift from the Old World. Cuitláhuac reigned for only four months before succumbing to the terrible disease. Montezuma's 22-year-old nephew, Cuauhtémoc, replaced him.

Ten months later, Cortés returned and lay siege to the city in May 1521. He had about 850 Spanish infantry, eighty-six cavalry and fifteen cannon. The infantry included 120–160 crossbowmen and arquebusiers. Artillery included three heavy iron cannon and five small bronze ones, as well as 510kg (1,124lb) of gunpowder. Perhaps more significantly, his Tlaxcalan allies provided an army of up to 24,000 men.

After the flight from Tenochtitlan, Cortés had asked after Martin López, not because he wanted the ships repaired for a retreat to Cuba, but because he wanted a fleet to sail upon Lake Texcoco. On the coast Cortés had thirteen brigantines (small ships) built which could carry cannon in their bows. They were moved inland by up to 8,000 Indian porters and escorted by Diego de Sandoval with 20,000 warriors.

The Aztec island fortress was well defended. The three causeways were actually dykes with breaches to allow the water through. These breaches were crossed by bridges which, as the Spaniards knew to their cost, could be removed and were protected by guard towers. Cuauhtémoc would have been well informed of the conquistadors' preparations and would have called in supplies and men from the surrounding region.

Indeed, Cortés knew that to reduce Tenochtitlan he must first cut it off from all outside help, both manpower and supplies, and then starve its already weakened population. He instigated a policy of divide and rule amongst the tribes, using the sword and diplomacy. His forces moved to quickly reduce all the cities and towns around Lake Texcoco and the other adjoining lakes. They also prevented the Aztecs from gathering their crops on the shores of the lake by intercepting harvesting parties.

Cortés divided his forces into three divisions, under Pedro de Alvarado (200 Spanish, three cannon and 30,000 allies), Cristóbal de Olid (220 Spanish, two cannon and 30,000 allies) and Diego de Sandoval (200 Spanish, two cannon and 40,000 allies).

Despite Cuauhtémoc's efforts, the old Triple Alliance crumbled. In particular, a major blow to Tenochtitlan's defence was the defection of the city of Texcoco. Lacking the island defences of Tenochtitlan, the Prince of Texcoco felt it prudent to side with Cortés when he arrived. The Texcocans had tasted Spanish steel at Otumba and wanted no more of it. This gave the Spanish a base near the lake and additional manpower. Texcoco was fortified and 8,000 workers built a canal linking the city with the lake ready to launch the brigantines. The Texcocans also provided the Spaniards with valuable munitions. They copied Spanish crossbow bolts and bolt heads, producing 50,000 bolt shafts and copper bolt heads within eight days. These were said to be superior to the Spanish supplied ones.

Aztec towns and cities were not fortified; only the fortunate ones were built on islands, which made them difficult to reduce.

Spanish tactics were to follow a similar pattern, several hundred Spaniards with several thousand allies would approach a town, the garrison would issue forth, be defeated and the town sacked. Just a week after their arrival at Texcoco, one of the first cities to feel the Spaniards' vengeful wrath was Iztapalapan.

Lying on an isthmus to the south of the Aztec capital, it was home for upwards of 50,000 inhabitants. Cortés marched on the city with 200 Spanish infantry, eighteen cavalry and 3,000 Tlaxcalans. An Aztec force advanced to meet them but was driven back and the city fell with the loss of 6,000 men, women and children. As Iztapalapan burned, the Aztecs breached the nearby dykes and almost drowned the invaders. West of Tenochtitlan at Tacuba (Tlacopan), the defenders were also brushed aside. The city was again burned and the Spaniards pursuing the fleeing people across Tenochtitlan's causeways were nearly trapped. Cortés stayed in Tacuba for six days then returned to his base at Texcoco, having been away for nearly two weeks.

The Spaniards' terror tactics worked, and soon deputations from around the valley and beyond were coming to Cortés, including the rulers of Otumba. In the face of all the destruction ten local kings sided with the invaders. The city of Chalco on the far eastern shore of the lake surrendered, despite the presence of an Aztec garrison. It is believed that Cortés may have ended up with up to 60,000–75,000 Indian allies camped round Lake Texcoco. Gómara, Cortés' secretary in later life, claimed they had 200,000 allies at the siege.

A Spanish force under Sandoval of 300 infantry and twenty cavalry was despatched to Chalco. The local Aztec forces were defeated and their nearby stronghold reduced. Afterwards Cortés occupied and sacked Xochimilco. He drove off the Aztec warriors, but was almost captured and only just saved by the intervention of a Tlaxcalan.

It seems the Spaniards did not bother with the ancient city of Tula, to the north of Lake Texcoco. Once the capital of the Toltec civilisation, it was still the home to perhaps 20,000 people. Cortés' spies would have informed him that it was no longer a seat of power. The Aztecs had long since looted the Tula Grande pyramid complex and many objects had ended up in Tenochtitlan and other Aztec cities. Nor did they bother themselves with Teotihuacán, which lay to the north of the city of Texcoco. This was formerly another mighty Toltec city which, in its day, had rivalled Tenochtitlan in population terms, with some 75,000–200,000 inhabitants. Despite its massive temples, it was now little more than a village. The city's original name is lost, but in Aztec it means 'the abode of the gods' and Montezuma and his priests still offered sacrifices there.

Having successfully isolated Tenochtitlan from its allies, Cortés' next move was to cut the city off completely. Once the brigantines were ready, they sailed out to wrest control of the lake and the fresh water supplies were severed. The brigantines were also used to protect the attackers' flanks as they advanced over the causeways. The boats' cannon were then deployed to raze the walls and buildings to allow the cavalry to manoeuvre once ashore. Aztec weapons were no match for the Spaniards' arms and armour, though the climate rapidly rendered crossbows and arquebuses useless. Their javelins and arrows seemed only capable of inflicting wounds. However, the Spaniards feared the Aztec sling, which could deliver a skull-shattering missile.

The Spaniards soon gained a lodgement in the southern part of the city and fought their way north, reportedly losing 2,000 Indian allies in one battle. The Aztecs, having decided to fight to the death, forced the Spaniards to slowly raze Tenochtitlan to the ground. In response, the Aztecs sought desperately to appease the gods by sacrificing every Spaniard they could lay their hands on. Even horses' bloody heads were displayed on the *tzompantli*.

Scores of Spaniards were captured and this had an impact on the prosecution of the siege. Spanish tactics were to send forward the Indians to clear the Aztecs' defences, withdraw them when counter-attacked and then bring up the Spanish cavalry and foot. During the night, the Aztecs would sally forth to break down the bridges placed across the canals in the day. The population, in the meantime, was stricken with starvation, colds, smallpox and measles.

The Aztecs learned tactics of survival against the Spaniards' missiles. In response to the crossbowmen taking aim they ducked, while to avoid the lead balls of the arquebusiers they zigzagged. They also overcame their fear of the roar of the cannon and knew to throw them in the lake if they captured one. In response to the Spaniards' cavalry, they knocked gaps in the houses as boltholes from which they could issue after the retreating horsemen.

At one point, it seemed as if the exhausted defenders received valuable reinforcements from the Xochimilcans. These people lived on the island in the south of Lake Texcoco and, slipping through the Spanish boats, reached the Aztecs. Cuauhtémoc rewarded them with gifts, but come nightfall the Xochimilcans tried to drag off Aztec women and children. Enraged, the Aztecs fell upon them. The Spaniards, in turn, burned their city.

Day by day, the defenders grew weaker and were reduced to eating bark and grass. Finally, the entire southern end of Tenochtitlan was conquered, leaving the population hemmed up in Tlatelolco. The Tlatelolco warriors had always claimed greater prowess than their southern neighbours, now they were able to prove it. Before taking the Tlatelolco market place, the Spaniards were almost trapped by them and had to retreat across the causeway to Tacuba. Morale was so poor in Cortés' army that many Cholulans, Texcocans and Tlaxcalan warriors began to slip away and had to be rallied. Cortés found his Indian forces were only a fraction of what they had been and his gunpowder ran out. Even so, the conquistadors' war of destruction continued to prevail.

Once the Spanish had a foothold in Tlatelolco, Alvarado's Tlaxcalan allies launched an attack into the city unassisted for the first time. When the time came to withdraw, a 400-strong Tlaxcalan rearguard of archers halted the Aztec pursuit dead in its tracks. The final battle for the great market place lasted five days and ended with 2,000 Tlatelolcan dead. Gradually the fighting was confined to the north-eastern suburb of Yacacolco.

Then, on 13 August 1521 the unfortunate Cuauhtémoc was captured and resistance collapsed. He and his family had taken to a canoe, perhaps with a view to escaping or surrendering. Either way, Cuauhtémoc's fate was sealed, as was Tenochtitlan's. The once proud city, which had held out for eighty days, lay in ruins with its surviving population starving to death.

The siege cost perhaps up to 100,000 Indian dead; in stark contrast, the Spaniards lost fifty men and six horses. It is claimed that of the 300,000 warriors defending Tenochtitlan only 60,000 were left. Allowing for contemporary embellishment, these figures probably reflect the total Aztec population. The Spaniards destroyed most of the buildings and, in particular, pulled down the sacred areas so encrusted with dried human blood. Cortés razed the temple of Huitzilopochtli and built a cathedral on the grounds.

In the wake of the fall of Tenochtitlan, King Cazoni of the Michoacán submitted to the Spanish, but Sandoval had to fight a pacification campaign elsewhere. Cortés' reward for bringing down the Aztec Empire was to be appointed governor and captain of New Spain. Cuauhtémoc was to be eventually murdered after the Spaniards had exhausted their search for elusive Aztec gold.

It was estimated in the 1950s that the value of gold Cortés took from the Aztecs was worth about $6.3 million, a king's ransom by anyone's reckoning. It has also been estimated that forty years after the invasion about 101 metric tonnes (99 tons) of gold had been shipped from the New World to Spain. The conquistadors' quest for gold was far exceeded by the quantities of silver they found. Between 1503 and 1660 16 million kilograms of silver were shipped to Seville from the New World; in contrast, only 185,000 kg of gold were shipped (which did, however, increase Europe's supplies by one-fifth). The lure of treasure ensured that by 1570 there were 118,000 colonists in the New World, sealing the fate of the pre-Columbian Indians.

Having conquered the Aztecs, Cortés turned his attentions on the Maya and the Yucatán Peninsula. Alvarado, butcherer of Cholula and Tenochtitlan, was set upon them. In 1523 he began conquering the southern Maya in Guatemala, subduing the Cakchiquel, Quiche and Tzutuhil Indians. He burned Utatlán, the Quiche capital, and was appointed captain general, though resistance continued until 1541.

Cortés then sent an army against the northern Maya in 1527. They were expert jungle guerrillas and it was not until 1542 that they were completely defeated. Ten years after the fall of Tenochtitlan, the Inca Empire in Peru was to suffer the same fate. This was five times the size of modern France, making it vastly bigger than the Aztec Empire and much more difficult to bring to its knees.

The Spanish brought the Aztec Empire down through a mixture of audacity and guile. The Aztecs could never adjust to the Europeans' style of warfare or match their superior tactics and weapons. The Indian mass attack was all but useless, as long as the Spaniards kept their heads the Aztec weight of numbers could not be brought to bear. They fought with different weapons and totally opposed concepts of war.

The Aztecs found themselves outclassed, even by such small numbers, and betrayed by their vassals. The Tlaxcalans and Texcocans played no small part in the destruction of Tenochtitlan. Aztec architecture also contributed to their defeat. Their towns were not fortified, which made them difficult to defend; only Tenochtitlan, built upon the lake, was able to hold out for any length of time.

Cortés may have been an inspirational leader, but it was not just his strategy, tactics and numerous Indian allies that gave him victory at Tenochtitlan. Smallpox and other European diseases take some credit for fatally weakening the population before the siege had even commenced. While his army was unable to prevent its ejection from the city, smallpox certainly helped him get back in. Tenochtitlan was ultimately a victim of its own success. Resented by its weaker neighbours, Cortés was the catalyst that finally galvanised resistance and finally orchestrated the overthrow of Tenochtitlan's power. It was a triumph of the Old World over the New World and a taste of things to come in Europe.

7

BASTION OF CHRISTENDOM: ST ELMO 1565

The Mediterranean was beset by a major crisis in 1565. The sultan, Suleiman the Magnificent, ruler of the Muslim Ottoman Empire, decided to direct his military efforts against Sicily and Italy, but first he required a safe naval haven from which to launch his attacks. The island of Malta was chosen as that base. The sultan, addressing his generals said, 'It is my greatest desire to capture Malta, not for its own sake, but because of other and greater enterprises which will follow this expedition.' He planned to raid the western Mediterranean, take Sicily, Hungary and the Holy Roman Empire.

The siege of Malta is one of the most renowned defensive actions in the history of siege warfare. For four months, a force less than one-third of the size of its attacker held out with the aid of human ingenuity, gunpowder, cold steel and the zeal of Christianity. Militarily, it heralded the high watermark of the Ottoman Empire in the Mediterranean and reversed the trend of the previous century whereby artillery had triumphed over medieval fortifications. The development of firearms in the early fourteenth century had transformed not only warfare but also individual soldiers themselves. Innovative Ottoman tactics had put them ahead of their opponents. The Turks relied primarily on firepower, while the Europeans relied mainly on shock. Throughout Europe, artillery had been used in siege warfare for over 100 years, but by 1565 the Ottoman Turks were considered masters of this art: especially after their reduction of Constantinople in 1453.

In the sixteenth century, the Turks were largely unique in sustaining sizeable regular armies, which were both paid and well disciplined. Ironically, not only were they slaves, but they were also non-Muslims recruited through '*devshirme*', a tribute of male children levied on the empire's Christian subjects. The Turks also initially had a technological lead, as they embraced the large-scale use of firearms sooner than most.

When the last Christian stronghold of Acre fell in Palestine in 1291, the Hospitallers Order of St John found themselves homeless for the first, but not the last, time. They remained exiled in Cyprus until 1310 when they became custodians of Rhodes. Only 12 miles from the Ottoman mainland, Rhodes was very vulnerable to attack. This close proximity of infidels was seen as an affront to Islam. Indeed, in 1480 about 450 members of the order and 4,000 troops found themselves successfully resisting an invasion of 70,000 Turks. The siege lasted eighty-nine brutal days and Muslim losses were estimated at 9,000 dead and 30,000 wounded. Not surprisingly, for over forty years the Turks left the island in peace.

The growth of the Ottoman Empire had been based on the fiery impetus of the Islamic faith. Expanding both east and west, it dominated North Africa, Mesopotamia and the Middle East and finally, with the fall of Constantinople, began to push into eastern Europe via the Balkans. The sixteenth century saw the Ottoman Empire at its height but also witnessed the beginning of its slow decline. By 1517, Egypt had become part of the Ottoman Empire along with the Levant ports, completing the Turks domination of the entire eastern Mediterranean.

The principal danger to Christendom stemmed from eastern and central Europe's inability to hold the Ottomans at bay. The line of the River Danube protected eastern Europe until 1521 when Belgrade fell. The Turks then moved north into Hungary, defeating the Hungarians at Mohacs in 1526. Vienna was besieged three years later, although the siege was eventually abandoned and was the high watermark of Turkish expansion in Europe. After Belgrade, Ottoman aspirations were directed at Spain and Sicily, resulting in attempted domination of the western Mediterranean.

Spain was the champion against the Ottoman menace. While the Ottomans posed little direct danger to much of western Europe, for Spain the threat was very real, Granada, in southern Spain, faced the North African Muslim states of Morocco and Algeria along the Barbary Coast. The semi-autonomous North African states were corsairs and slavers by trade, continually harassing the Spanish coastline and Mediterranean commerce.

To France, the immediate threat was Spain; as a result, her foreign policy tended to be pro-Ottoman, although this was, of course, never announced publicly. During the sixteenth century, France allowed Ottoman galleys to use southern French ports, from where they could harass Spanish shipping.

Rhodes was the first Christian Mediterranean island to be subjected to the finely honed power of the Turkish siege machine. In 1522, fresh from his success at Belgrade, Suleiman the Magnificent collected an army of up to 140,000 men and 400 vessels – just 600 knights and 4,500 troops defended

Rhodes. The invaders arrived on 26 June and spent thirteen days unloading troops and supplies. Initially the siege did not go well for the Turks, and Suleiman himself arrived on 28 July with 15,000 fresh troops.

The defence of the Bastion of England stands out, for it became the main objective of the Turks. Two mines were blown under the bastion on 4 September 1522, bringing down 12 yards of rampart. This was followed by an assault that left 3,000 attackers dead. Denied reinforcements due to bad weather, the defence of Rhodes become untenable and, after six long months, on 20 December the Grand Master reluctantly capitulated. Just 180 knights and 1,500 men marched out with full military honours. In total, the Turks fired 85,000 iron and stone shot, dug fifty-four mine tunnels and staged twenty assaults during the siege. An impressed Charles V of Spain remarked, 'There had been nothing so well lost in the world as Rhodes.'

The knights were homeless again until 1530 when their galleys arrived in Grand Harbour, Malta, a gift from Charles V. They were also given responsibility for Tripoli on the North African coast. Spain had been busy securing the North African ports, having captured Oran and Mers-el-Kébir in 1509 and Tripoli the following year. In 1535, the order assisted Charles V's forces in capturing Tunis. He then led a disastrous expedition against Algiers in 1541, which was the sole remaining Turkish garrison along the Barbary Coast.

Ten years later, the Turks fell briefly on Malta in 1551 with 12,000 men, although their intended and subsequent target was Tripoli, which indeed capitulated. The order lost nineteen bronze cannon at Tripoli, which had been a gift from Henry VIII. The port's lack of preparation and the small size of its garrison were important lessons learned by the order. This loss and constant clashes with the corsairs and the Turks

ensured the knights did all they could to improve the fortifications on Malta.

The island had far greater strategic significance than
Rhodes did, which had largely been a distraction for the
Turks. Malta formed a bottleneck between Sicily, Italy and
the North African coast and provided a key defensive point
for the western Mediterranean. After their successes at
Rhodes and Tripoli, the Turks did not take into account the
resolution of the Knights of the Order of St John to defend
Malta. The Grand Master Jean Parisot de la Valette was not

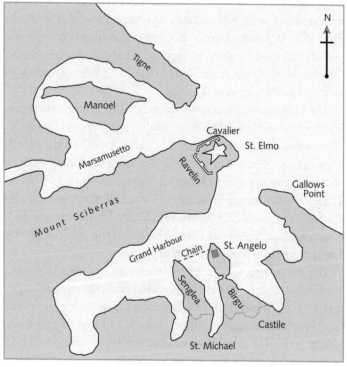

Defences of Grand Harbour, Malta 1565.

only aware of the danger posed to southern Europe if Malta should fall, but also the shame the order would have to endure. The knights' heroic defence of Rhodes had not saved the island, and once again the Muslim Turks were threatening to destroy their beloved order.

Malta was, in effect, a dependency of neighbouring Spanish Sicily and looked to the island for help; Sicily in turn looked to Spain. However, Sicily was tied down by the fear that France would side with the Ottomans. Certainly, the French and the Turks had long co-operated to counter Spanish ambitions. Malta could provide an ideal base from which the Turks could act with the French fleet at Marseilles and Toulon against Spanish interests. Indeed, the secret Treaty of La Forest, between King Francis I of France and Suleiman, had been renewed by Henry II of France.

The end result was that the Viceroy of Sicily had to look to his own defences, and in vain requested 25,000 soldiers from King Philip II of Spain. The order's admiring guardian, Charles V, had abdicated in 1555 and responsibility now fell to his son, Philip. He took the Turkish threat seriously. In early 1565 Naples and Sicily were put on alert and Spain's Italian possessions were instructed to raise 10,000 men. In reality, Philip did not have the resources to counter the magnitude of the Turkish military effort, or fight on two fronts, as trouble was brewing in the Spanish Low Countries (Belgium and the Netherlands).

A Turkish Army embarked at Constantinople on 29 March 1565. The fleet set sail in early April and on 18 May was sighted 15 miles from Malta approaching east-north-east. Compared to the forces the knights could muster, the Muslim force was huge. Francesco Balbi de Correggio, who served in the garrison of Senglea through the whole siege, gives the Turkish Army's composition as follows:

Number	Type
3,500	Iayalars (religious fanatics) from the Balkans
400	Iayalars from the Islands
6,300	Janissaries (elite arquebus armed infantry)
6,000	Levied troops (irregular infantry)
6,000	Spahis (light cavalry) from Anatolia
500	Spahis from Karamania
400	Spahis from Mitylene
2,500	Spahis from Roumania
4,000	Volunteers (renegades from Greece and the Levant)

The total number that sailed from Constantinople was 30,000, but including naval troops, Moors of Dragut, Pasha of Tripoli and Hassan of Algiers, the total number of fighters, not including sailors, was in the region of 48,000 men. Therefore, total Turkish forces investing the island, before illness and casualties reduced them, was at least 40,000.

At Rhodes, the Turks had employed thirty-three guns and twelve bronze mortars capable of firing explosive rounds. For the siege of Malta, the Turks had twice the number of artillery, with some eighty guns (including twelve mortars, probably the same weapons) ranging in calibre from up to ten 80-pounders and one massive 160-pounder. The Turkish gunners were more than confident that they could reduce Malta's impressive defensive stonework.

Perhaps the most significant of the invading force were the 4,500 Janissary arquebusiers, for they easily outclassed their Christian counterparts. The most famous of the Turkish forces, the Janissaries (*Yeniceri*), or 'New Soldiers', distinctive in their white sleeve-cap headgear or *zarcola*, were formed in 1362. They were second only to the Spahis of the Porte as the core of Ottoman military power. Initially raised from prisoners of war, they were also raised by *devshirme*.

By the sixteenth century, they were armed with a firearm
called the arquebus.

Such dedicated firearm infantry, particularly outside
Europe, gave the Turks a distinct advantage over their eastern
neighbours such as the Mameluks and Persians who often had
better cavalry. 'Spahis' means soldiers, but in the Turkish context they were cavalry. They comprised two types – the paid
regular Spahis of the Porte (household cavalry) and the more
numerous feudal cavalry. Under Suleiman the Magnificent,
the Spahis of the Porte had expanded to about 12,000.

In stark contrast, the Knights of the Order of St John
could muster less than 10,000 men, who were a hodge-
podge of different nationalities from across Europe and the
Mediterranean:

Number	Type
500	Knights of all *langues*
400	Spaniards of the companies of Miranda and Juan de la Cerda
200	Italians of Asdrubale de' Medici
400	Under Colonel Pierre de Massuez Vercoyran (Mas)
200	Company of de la Motte
100	Soldiers of the ordinary garrison of St Elmo
500	Soldiers of the galleys
100	Servants of the Grand Master and Knights
200	Greeks and Sicilians
500	Galley slaves and hired oarsmen
3,000	Maltese

The garrison stood at 6,100 men, half of whom were Maltese,
but was boosted by reinforcements to reach just over 9,000.

The Turkish fleet was initially anchored off Mgarr, but
on 19 May it moved to Marsa Scirocco. That day saw the

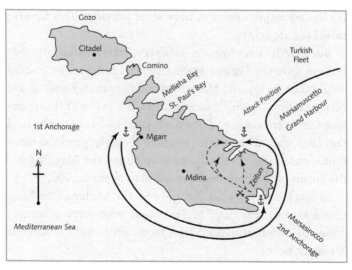

Ottoman invasion of Malta, 1565.

first skirmish fought at Zeitun. The Turkish commanders, Mustapha Pasha and Admiral Piali, although divided in opinion, decided to first secure Marsamuscetto Harbour, just north of the Grand Harbour. Vitally for the defenders, Piali was mistaken when he thought Marsa Scirocco was not a safe shelter for the fleet. In order to secure Marsamuscetto, the small fort of St Elmo, which dominated both the entrances to Marsamuscetto and the Grand Harbour, would have to be captured.

St Elmo consisted of a star fort, with four salients shielded by a defensive outwork, or ravelin, on the landward side to the south-west, and a tower and a raised gun platform, or cavalier, to the north-west facing the sea. After a survey by Turkish engineers, it was optimistically concluded the fort would last at the most five days. Opposite, across Grand Harbour, built on a strip of land stood the town of Birgu with the Castle of

St Angelo defending its habour side. To the south, separated from it by the Harbour of the Galleys, was the new town of Senglea, which was protected on its landward side by the Fort of St Michael opposite Birgu's Castile fortification.

It was formally decided to attack St Elmo on 22 May, and doing so gave Grand Master Valette a breathing space to continue preparations and the Governor of Sicily time to raise a relief force. Three days later, the heavy guns from the Turkish fleet were transported to St Elmo – a distance of some 9 miles, it required a dozen bullocks to pull each gun. Valette sent 100 knights, Colonel Mas and La Motte with their companies to St Elmo, boosting the garrison to 800 defenders.

By the 26th the Turkish approach trenches were 600 paces from the fort's ditch on the Marsamuscetto side of the point. The approach from the south side could not be seen from St Elmo and the Turks soon reached the counterscarp of the ditch. Earth-filled wicker gabions were erected on the Marsamuscetto side, along with gun batteries. The following night, the artillery opened up along with Turkish arquebusiers who, if they failed to fire, were threatened with 100 lashes on the stomach.

By the 29th, the Turks were keeping up a concentrated fire, although counter-fire from the fort knocked out some of the besieging guns. However, the following day a large culverin had to be brought down from the Tower of Sagra, because the structure was unsafe. Some twenty-four pieces of Turkish artillery then opened up a murderous bombardment on 31 May.

It was not until 2 June that Rias Dragut, corsair and Governor of Tripoli arrived. He had been appointed by the sultan as advisor and mediator between Mustapha and Piali. Dragut joined the fleet with thirteen galleys, two galiots (small ships), thirteen other pirate vessels and 2,500 volunteers. He was displeased that the neighbouring island of Gozo and northern Malta had not been secured first, thereby sealing communication with Sicily, but fatally decided that what had been started must be first finished.

A gun platform was constructed that could bombard the Tower of Sagra and the ravelin. Then, the day after Dragut's arrival, a surprise attack resulted in the capture of the ravelin. The guards were probably shot dead by snipers and, undetected, the Janissaries swarmed up the walls. Most of the defenders were massacred and the Janissaries stormed across the bridge between the ravelin and St Elmo. Two cannon on top of the gatehouse blasted the Turks back as the few survivors fled into the fort. Some fifty Christian troops also withdrew to the trench between the ravelin and the tower, but they were forced out.

The garrison launched a counter-attack and the struggle lasted for five hours. When the two sides finally parted, nearly 2,000 Turks had been killed and wounded, along with sixty Christian soldiers and twenty knights. Crucially though, the Muslims remained in control of the ravelin, and its loss spelled disaster for the defenders of St Elmo. Using earth-filled goatskins, the ravelin was raised to the same height as the fort's walls. The Turks then mounted a platform with two guns on it, thus enabling them to shoot into the interior of the fort.

Turkish troops entered the ditch between St Elmo and the ravelin on 4 June and began to construct a bridge opposite the post of Colonel Mas. Christian gunfire succeeded in forcing them off. Undeterred, the Turks then removed stones from the wall and brought up ladders, constructing another bridge below the ravelin. The defenders, in turn, sallied out and burned three of the bridge's five supports. However, all this was in vain, for on the next day the Turks built a third bridge. With it firmly in place, the Muslims launched a general assault over it on the 8th. The garrison placed earth-filled gabions on their end of the bridge and, after seven hours, the Turks were again beaten off with the loss of 500 men, while the knights lost forty.

Two days later, the Muslims began to fill in the ditch and raised ladders near the old bridge. The aggressive knights sallied out again and attacked the workers. Both sides fed troops into the struggle and the fighting lasted until dawn. The Turks lost another 1,000 men; the defenders sixty, 'the greatest number of casualties being caused by our fire hoops', noted one of the defenders with satisfaction.

On 11 June, 150 reinforcements were sent to aid the exhausted garrison. For the defenders there was to be little rest, from 11–14 June the Turks sustained a terrible continuous bombardment. The battered fort was once more assaulted on the 15th and the struggle lasted for four hours. Another attack in the afternoon was also beaten off. The Muslims suffered 1,400 casualties and the Christians a further sixty.

The Turks, still undeterred by their terrible losses, planned yet another general assault for the following day. They gathered on the neighbouring high ground and spent two hours before daybreak praying. The priests absolved them of their sins and exhorted them to fight and die for their faith. Then, some 14,000 Muslim arquebusiers formed an arch across the peninsula from the edge of Marsamuscetto to the Grand Harbour. A murderous bombardment followed, supported by Piali's fleet, and St Elmo disappeared in a vast cloud of smoke and dust. The stone fortifications were pounded to rubble as the air filled with spent gunpowder. The Turkish Army marched forward led by the fanatical Iayalars. They were met as usual by a hail of cannon fire, arquebus fire, fire hoops, fire bombs, boiling pitch and masonry. Also, a battery on the southern side of the fort was able very effectively to enfilade the advancing Turks.

While the attackers swarmed around the walls, thirty Turks managed to gain a footing by means of steps they had cut in the rock on the north-west of the cavalier towards Marsamuscetto. This was on the post of Colonel Mas. The gunners in St Angelo saw the danger, and their artillery began to sweep shot along the exposed cavalier. Unfortunately, they killed several of St Elmo's defenders, but corrected their aim and twenty of the Turks were cut down. The rest (mostly officers) were killed by the garrison and thrown from the walls.

The Iayalars were also beaten back, as were the Dervishes, or Moors, so the Janissaries were finally committed to the assault. The nightmare struggle raged on for seven hours. The Turks continually attacked, regrouped and attacked again, but each time they were driven off.

By now, the exhausted defenders were on their last legs and could not have withstood another assault. A Captain Medran rushed forward to remove a Turk planting a flag on a gabion, only to be shot down. Another wounded knight heroically fought from a chair. When the Muslims eventually withdrew they left 1,000 casualties to the loss of 159 defenders. The Turks also lost three standards; one of which was Pasha's and another Dragut's. They could only lick their wounds and count their casualties, who numbered over 4,000 men in just three weeks. During the night a further 150 reinforcements were sent over to St Elmo.

The Turks, using six pieces of their main battery, bombarded the spur to the south on 17 June. The following day, two more guns added their firepower. Dragut also posted a battery commanding the cave below St Elmo, the fort's secret harbour entrance. The Turks then built a ditch and wall from the counterscarp of the fort's ditch to the shore facing St Angelo. All this effectively sealed St Elmo off from the outside world.

On the 18th, the grand master was presented with the news, brought by a Lombard renegade, that Dragut was dead. De la Valette observed, 'which news was welcome to all and that he [the Lombard] had seen him prostrate with his brains protruding from his mouth, nostrils and ears.' Dragut had been in the counterscarp in a double trench when his gunners were firing too high; he ordered them to lower their aim. He had his back to the guns to observe their fire when a shot struck the trench behind him and a splinter of rock caught him in the head (other versions of events have the shot falling from St Angelo and St Elmo).

The loss of this expert soldier was a dire blow to the Turks' efforts. They directed a total of ten cannon against the southern spur, which had been blasted almost level by 19 June. The defenders were desperately using mattresses, blankets and wetted sails against the Turks' incendiary devices. Also, the battery along the coast had effectively stopped the defenders reaching the Grand Harbour. The following day, the battery dominating the cave was completed, finally cutting St Elmo off from all external help. The fort's ditch was slowly filled up and the bombardment continued. St Elmo's gunners tried unsuccessfully to shoot the Muslims' bridge away. A volunteer even slipped over the wall and tried to set fire to it, but the bridge had been covered in damp soil. By now, the Turks had completed a breastwork on the captured cavalier.

The next general assault was launched on Friday, 22 June. The hottest fighting took place on the bridge and around the bastion of Colonel Mas. Once across the bridge, the Turks tried to attach ropes to the defenders' gabions in an attempt to pull them down. Once again, the Turks on the exposed salient were shot at from St Angelo. The knights, as usual, presented their wall of steel along the walls and the Muslims, for all their efforts, could not enter. The defence lasted for six hours, after which the Turks could endure the carnage no longer and retired. Around the walls they left a further 2,000 men, but this meant little to the defenders who had been crippled beyond hope. The knights had lost 500 men and all their officers, leaving only about 100 able-bodied defenders.

That night, a swimmer reached the grand master. On hearing of the plight of the defenders, Valette's resolution to sacrifice the fort flagged and he tried vainly to send a relief force across the harbour. The reinforcements not only came under cannon fire, but found enemy boats waiting for them and had to withdraw. The end for St Elmo was now in sight.

The Turks prepared for their final onslaught on 23 June, appropriately the eve of the Feast of St John. The defenders managed to brave the storm for about four hours and miraculously the attackers could still not gain entrance. Nonetheless, the Turkish arquebusiers occupying the cavalier and other high points were able to fire on the remnants of the garrison, now some sixty wounded men.

A parley was called for, but the Janissaries entered by the Sagra and threw stones at the men defending the bridge to distract them. The Janissaries cried for the attack to continue and the Turks swarmed over the bridge and the salient of the spur. The defenders were finally swept away. Some retired to the church, hoping to surrender, but the maddened Turks beheaded those who did. The remainder rushed to the square to make a last stand. These gallant knights stood back to back and died as the honour of their order dictated. Those not shot were hacked and stabbed to death by their assailants. Five Maltese managed to swim to Birgu across the harbour, but those soldiers who remained alive were cut to pieces. Just nine wounded knights, in the guardhouse at the entrance to the ditch behind the church, surrendered to the corsairs. They hid them in the hope of ransom, although Mustapha claimed them.

The siege of St Elmo had lasted for thirty bitter days. For the Knights of the Order of St John the bloody struggle for this fort was over; for the Turks, the siege of Malta was only just beginning. The Turks expended about 18,000 rounds and allegedly lost 8,000 men, nearly a quarter of the force that had sailed from Constantinople. They captured twenty-eight cannon and one culverin. The Christian defenders lost 1,500 men, including 130 knights. The pasha was astonished when he went into St Elmo, and said, 'Allah, if the son which is so small has given so much trouble, what will the father do, which is so large?' Mustapha Pasha was, of course, right, for the siege of Malta was to break the back of the Turkish Army, ready for the relief force to arrive and deliver the *coup de grâce*.

In his rage, the pasha had the dead knights' heads mounted on stakes facing St Angelo. The bodies were mutilated with the sign of the cross scored across the chest and crucified. The wooden crosses were then tossed into the harbour where the current carried them to the shores of Birgu. In retaliation, the grand master had all his Muslim prisoners executed and their heads fired into the Turkish camp. From that point on, no quarter would be given or expected – it was a battle to the death.

It was not until the end of June that a small relief force of four galleys and 600 men arrived. They initially made their way to the walled city of Mdina in the middle of the island, and then on 5 July, using small boats, slipped past the Turks to reach Birgu.

After the fall of St Elmo, the Turks moved to invest Fort St Michael, south of Senglea, which along with St Angelo

and the town of Birgu formed a self-contained commu-
nity linked by a vital bridge. To facilitate this and avoid
the guns of St Angelo, the Turks hauled eighty ships from
Marsamuscetto Creek across Mount Sciberras to get them
into Grand Harbour, below Senglea. A week-long artillery
bombardment by sixty to seventy cannon was followed by
a full-scale assault. A concerted naval and land attack was
launched against St Michael and Senglea on 15 July. The
Turks were to lose another 3,000 dead and the defenders, 200.

Throughout August the Turks continued their bombardment and massed assaults, using all their ingenuity to level
the defiant fortifications. A general assault was launched on
7 August, when 8,000 men were thrown against St Michael
and another 3,000 against the Bastion of Castile, which protected the southern of side of Birgu. The Turks carried all
before them and Senglea was about to be lost when Mustapha
sounded the retreat. The Governor of Mdina in fact saved
Senglea. Realising the situation must be critical from all the
gunfire to the north, he sent out a mounted force to try to
draw off some of the attackers. In the confusion, the Turks
thought a relieving army had arrived from Sicily.

On 18 August 1565 the Turks exploded a mine underneath the Bastion of Castile and for a while it looked as if
the Turks might enter the town via the breach blasted in the
walls, but the defences held. Things had become so critical for
the defenders that three days later the grand master issued an
order that all walking wounded in the hospital must return to
the walls. It seemed that time was running out and the will of
Allah would prevail.

A relief force of twenty-eight ships with about 8,000 men
finally arrived on 7 September to confront some 10,000 sick,
wounded and demoralised Turks. Initially abandoning their
siege lines, it seemed the Turks were intent on withdrawing,

but they had a change of heart when they learned just how small the relief force was. Mustapha disembarked 9,000 troops at St Paul's Bay. The trapped knights, determined to avenge themselves, instigated an attack. However, their first act was to cross Grand Harbour and raise the flag of the order once more over the ruin of St Elmo. After some fighting, the dispirited Turks were driven back into St Paul's Bay and fled to their ships. The knights waded into the surf after them, killing as many as they could lay their hands on.

Seven days after its arrival, the relief force's fleet entered Grand Harbour to be greeted by the sight of the blackened ruin of St Elmo and the shattered ramparts of Birgu and Senglea. Shortly after, a further 4,000 reinforcements arrived and it was agreed that a similar number of Spanish would be left on Malta just in case the Turks should venture back.

Philip II of Spain was criticised for not doing more for the beleaguered island. Nevertheless, to oppose the Turkish fleet and army at Malta too soon would have been madness. Spain did not have military resources that it could afford to squander. Just two years after Malta, Philip was forced to commit troops from Spain, Naples, Lombardy and Sicily to counter Flemish rebels in the Low Countries. The best the Spanish could do was get 1,000 men onto the island in June, followed by 8,000 in September, the latter arriving, by good fortune rather than judgement, in the nick of time to raise the siege.

Malta was to contribute to the final blow to the Turks' maritime pride in 1571 when galleys from the island helped defeat the Turkish navy at Lepanto. Ottoman sea power saw the loss of up to 100 ships and 20,000 dead or captured, proving to Europe that if they buried their differences, the invincible Turks could be defeated. The price of victory at Lepanto was 8,000 killed and nearly 16,000 wounded. Even so, the Ottoman threat still remained – despite their military setbacks, Lepanto proved to be a defeat for Turkish morale more than anything else. The Christian allies failed to follow up their victory and, as a result, the Turks simply rebuilt their fleet.

Ottoman success had been due to them having the only professional standing army, but by the seventeenth century this was no longer the case. The Turkish threat during the seventeenth century was overshadowed by the Thirty Years War, the English Civil Wars, the Dutch Wars and the Spanish War of Succession. Both France and Germany learned many important lessons, and large numbers of trained and experienced troops became available in Europe. By the 1640s European armies were combining firepower, shock and flexibility. The Ottomans, in contrast, remained entrenched in their old ways, employing thousands of irregulars.

The Ottoman Empire was beset by many problems, which ultimately undermined its ability to threaten Europe, again most notably becoming over-stretched. An empire built by force can only be maintained by force, once the expansionist momentum begins to slow, stagnation and decline is assured. The campaigning army has to wage constant war in order to maintain the expansion and security of its frontiers, creating a continual economic and manpower drain.

In 1664 at St Gotthard, the Ottoman Army was defeated for the first time on European soil. Poland then successfully defeated the Turks at Chotin in 1673. Ten years later, 250,000 Turks invaded Hungary seeking revenge for their failure in 1529. On 12 September 1683, the Turkish Army outside Vienna was completely smashed, followed by defeats at the Second Battle of Mohacs in 1687, Salankemen in 1691 and Zenta in 1697.

Nonetheless, it was not until 1918, with the end of the First World War, that the Ottoman Empire finally collapsed, although from 1688 onwards it never again posed a serious threat to Europe. St Elmo had first shown Europe the way ahead.

In 1565 the Order of St John lost almost 250 knights, while the other defenders and the inhabitants suffered 7,000 dead. By the time the siege was lifted, of the original garrison of 9,000, the grand master had about 600 still capable of bearing arms. Clearly, they were on their last legs and help arrived just in time.

Turkish losses are estimated between 20,000 and 30,000; Balbi claims the higher figure. Minimum losses of 20,000 Turks do not take into account Algerian, Egyptian and Corsair casualties, so the higher figure would not seem unreasonable. Turkish gunners fired in excess of 70,000 cannon balls at the battered defences of Grand Harbour. Despite their skill and experience in this field, the Turkish artillery had not been a decisive weapon. In fact, it was simple attrition and overwhelming force that had nearly brought them victory. In the case of Rhodes, it was mining rather than artillery that had won the day. Suleiman was dismayed, and recalling his victory at Rhodes, remarked bitterly, 'it is only in my own hand that my sword is invincible.'

8

No Surrender, No Quarter: Magdeburg 1631

The bloody rape of the German city of Magdeburg is comparable to the destruction of Acre and Tenochtitlan. Its loss during the Thirty Years War of 1618–48 became a byword for infamy and served as a salutary lesson in what happens to a besieged city that refuses to capitulate. The siege opened the third phase of the war, pitting Catholic German prince against Protestant German prince. It also represented a pyrrhic victory for the Catholic Imperialist cause, which was subsequently to suffer a series of military setbacks at the hands of the Protestant Swedish Army.

Magdeburg's desolation was a direct result of Sweden's meddling in German politics; indeed, the outbreak of the war had a whirlpool effect on much of Europe. Bohemia's demand for Protestant rule, rather than the Catholic Holy Roman Emperor Ferdinand II, led to the 'Defenestration of Prague', the crowning of Frederick of the Palatinate and ultimately revolt against the empire (i.e. Germany). What followed was four phases of the Thirty Years War, witnessing a total of thirty-three major engagements and the desolation of great swathes of central Europe.

The key weapons of the foot soldier by the 1600s were the smoothbore matchlock musket and the pike. The pikemen defended the musketeers from enemy cavalry while they were loading. The cumbersome seventeenth-century musket weighed up to 7kg and the heavy barrel had to be supported by a fork called a fourquette. It had a range of about 300m.

The backbone of the field and siege artillery was the cannon cast from gun bronze (copper, tin and lead). Greatly varying calibre guns fired anything from 1–80lb balls, while howitzers managed 6–30lb; mortars could deliver up to a 260lb ball. Canister and case shot was also used as an anti-personnel weapon, consisting of canvas or leather bags filled with anything from nails to rocks.

Cavalry comprised the arquebusier and the armoured cuirassier. The latter was a hang-over from the knight. Both were armed with wheel-lock pistols, while the arquebusier also carried a wheel-lock arquebus or carbine (becoming carabiniers). There was a third type, who were the mounted infantry known as dragoons, who rode to battle but fought on foot.

Frederick's failure at the Battle of White Mountain in 1620 meant the loss of the Palatinate, which was completely overrun within three years. It was not until 1626 and the second phase of the war that Christian IV of Protestant Denmark proclaimed his support for Frederick and invaded northern Germany. In consequence, the Bohemian Imperialist General Albrecht von Wallenstein defeated the Danes at Lutter and the whole of the Danish mainland was occupied, until only the port of Stralsund remained, which was besieged in 1628. Wallenstein's reward was a dukedom in Mecklenburg.

With Sweden's main rival beaten into submission and large areas of the Baltic coast under Imperialist control, King Gustavus Adolphus of Sweden had good reason to feel insecure. What his chief minister, Axel Oxenstierna said many years later sums up the situation, 'If the Emperor had once got hold of Stralsund, the whole coast would have fallen to him, and here in Sweden we should never have enjoyed a minute's security.' Sweden's intervention forced Wallenstein to abandon the siege and in 1629 Denmark made peace at Lubeck, ending the second phase of the war. The Swedes gained foothold in Stralsund.

Ferdinand II then aggravated matters by issuing the Edict of Restitution under which church property within the German Protestant states was to be returned to the Catholic Church. Two archbishoprics, including the city of Magdeburg on the Elbe, and twelve bishoprics were to be restored. This move alienated the powerful Protestant Electors of Brandenburg and Saxony. Wallenstein laid siege to the largely Protestant Magdeburg when it refused to receive and accommodate an Imperial regiment. After seven months, Wallenstein settled for a fine and marched away.

Prince Christian William of Brandenburg, the deposed administrator of the city, who had been living in exile in Stockholm, returned the following year with a pledge of support from the Swedish king. He had in readiness a force of 3,500 men and a promise of as many more from Weimar. A secret alliance was concluded between Prince William, Sweden and Magdeburg's town council. He headed for his residential capital at Halle, only to return to the city upon reports of approaching Imperialist forces.

In the meantime, Ferdinand II began to give aid to Spain fighting in northern Italy against the French, while Cardinal Richelieu of France began to stir up the German Protestant princes against the emperor. Although Catholic France was unable to commit herself openly, Richelieu encouraged Sweden to intervene further to keep the Protestant cause alive. This would effectively tie down the forces of the Catholic League and possibly the Spanish Netherlands, all of whom posed a threat to French interests.

By the Treaty of Barwalde, signed in 1631, France defined the goals of a Swedish invasion of northern Germany, and agreed to supply a subsidy over five years, in return for protection of continental commerce. For their part, the Swedes would maintain 36,000 troops in Germany, protect the freedom of worship for Catholics throughout Germany and refrain from invading the Catholic Duchy of Bavaria. King Gustavus Adolphus, though, was no mercenary and had his own strategic reasons for invasion – in particular establishing a security buffer. Significantly, he found he had no military support from the major Protestant German states for his enterprise.

It was while John George, Elector of Saxony, procrastinated whether to oppose Imperial authority, that Magdeburg impetuously pledged its support for renewed Swedish intervention. However, without the aid of Saxony and Protestant Brandenburg, the city was in a vulnerable position. Meanwhile Imperialist General Johann von Tilly had been despatched north with the Army of the Catholic League, to deal with the invading Swedish, who landed at the mouth of the Oder Estuary on 4 July 1630. Notably the Swedes invaded before Barwalde was concluded.

Gustavus Adolphus marched south through the Duchy of Pomerania with 40,000 troops, of which about half were Swedes and Finns. In his Baltic lands and at home he could call on another 30,000. The emperor, who was preoccupied with Italy, only had 40,000 soldiers spread throughout the German principalities. The Swedes could not tolerate Pomeranian neutrality and quickly occupied their capital, Stettin. The Duke of Pomerania reluctantly provided 3,000 men. He informed the emperor that he had only yielded because of superior Swedish military strength. Gustavus Adolphus then moved west to secure Stralsund before returning to Stettin.

To compound the emperor's difficulties, he was forced to dismiss Wallenstein in September 1630. This move, ironically, was to placate the Catholic principalities led by Bavaria who feared the general's growing power. Wallenstein withdrew to his Bohemian lands to brood. A much-reduced Imperial Army was placed under Duke Maximilian of Bavaria and General Tilly.

A distinguished Swedish officer, Hessian Dietrich von Falkenberg was despatched by Gustavus to Magdeburg on 29 October 1630 to organise its defence and galvanise Administrator Prince William. He took control of William's troops and imposed Swedish discipline. The following month, Gottfried von Pappenheim, with a small Imperialist force moved to invest the rebellious city. Negotiations followed, and Pappenheim tried unsuccessfully to bribe Falkenberg to leave. Slowly but surely, Magdeburg began to feel an Imperialist noose tightening around it while its Protestant saviours dithered. Tilly appeared in the New Year and called on the city to return to the Imperial fold, but was drawn away to assist the beleaguered Imperialist strongholds of northern Mecklenburg, particularly Demmin and Greifswald, south of Stralsund. In February, Demmin fell to the Swedes.

Tilly sensibly hoped to place the bulk of his Imperialist Army between Gustavus Adolphus' forces to the east in Pomerania and the Swedish forces under Marshal Gustavus Horn in the Duchy of Mecklenburg to the west. If he could achieve this he would be able to defeat the Protestants piecemeal. Tilly struck decisively between Adolphus on the Oder and his base on the Baltic seaboard by pushing into Mecklenburg. In a foretaste of what was to happen to Magdeburg on 19 March 1631, Tilly's forces stormed Protestant Neubrandenburg, butchering the Swedish garrison of 3,000 men. It was anticipated that Tilly would move on Stettin or try to relieve besieged Greifswald, but instead moved south, presumably to join Pappenheim.

Gustavus Adolphus, still lacking German Protestant support, decided to attempt to draw Tilly away from the Baltic ports and Magdeburg by attacking the Imperialist city of Frankfort-on-the-Oder. Leaving Horn to continue the siege of Greifswald, Gustavus with 14,000 troops drove Wallenstein's old command up the Oder into Frankfort. The city, despite being Protestant, was taken by storm on 13 April 1631 and, of a garrison of 5,000, around half were massacred with the cry of 'Neubrandenburg quarter!'. A Scots mercenary officer named Monro recalled he 'did never see officers less obeyed'.

The city was also subjected to three hours of raping and pillaging. This diversionary operation did not work, for Tilly's troops remained steadfastly at Magdeburg. However, Gustavus now controlled 80 miles of the Oder and his progress through northern central Germany could only be expected to ease the pressure on the city.

Gustavus succeeded in uniting with Horn's forces and Tilly was in fact forced to retire towards the Elbe due to large losses. Unable to shift Gustavus, Tilly was ordered by Maximilian of Bavaria to join Pappenheim at Magdeburg. Once combined with Pappenheim's soldiers, their army numbered about 22,000 foot, 3,000 horse or cavalry and eighty-five heavy guns, with another 3,000 Imperial troops at Dessau.

Falkenberg and his garrison were now potentially in the eye of an Imperialist storm. The beleaguered city had a populace of some 35,000 souls, which probably included refugees. The garrison numbered only 3,000 regular soldiers, possibly boosted to about 8,000–10,000 by levied citizens. The inhabitants, though, were on the whole not supportive now that they had so clearly incurred the emperor's wrath and this antagonism resulted in Falkenberg's cavalry mutinying for a time.

The ancient city defences consisted of an old medieval wall braced by numerous towers and small stone forts. With the growing troubles and the arrival of Pappenheim, these defences had been sensibly strengthened with a number of earth and timber forts. Earth was the new shock absorber, capable of withstanding the impact of iron cannon balls, as it was well known that the stone walls would simply crack and shatter. Therefore, Magdeburg was generally well fortified, except on the riverside to the north and north-east. To the north, where the River Elbe bisected the city, it isolated the suburb of Neustadt, while to the north-east islets bridged the marsh caused by the river's floodwater. This covered all the lowland in the area and there were three main islands forming a triangle, the top of which faced the north-eastern gatehouse. Falkenberg had substantially strengthened it with two earthen redoubts which now protected the approaches to this main gatehouse.

Facing northern Magdeburg, the right-hand island nearest the dry high ground had an earth fort built on it, the *Zollschanze*, which consisted of a large central square with bastions on each corner mounting several cannon. It was linked to the next island north-west of it by a lone wooden causeway. This next island was dotted with several houses and had a small round stone fort on its northern corner. The fort was separated from the island by a small moat on the southern edge, while to the north were the floodwaters of the Elbe. A small wooden bridge connected the island and the fort on the western edge of the moat. On the northern edge of the fort, a long causeway linked it with the city's north-eastern outer gatehouse, which, along with a battery of cannon and one small stone fort to the left and two stone forts to the right, were all outside the city walls. The island connected to the city was also connected to another island on its western bank by a small wooden drawbridge. This island, like its neighbour, was topped by an earth and timber fort.

Clearly Neustadt, the *Zollschanze* and the north-eastern gatehouse were the key to Magdeburg. Indeed, the Imperialist staff quickly realised the weakness of the city's north-ern defences and set about them. Furthermore, Catholic spies within the city would have kept them well informed. Pappenheim succeeded in capturing the redoubts on the right bank of the Elbe and one or two on the left bank. There then followed a breathing space after the news of the fall of Frankfort.

Fearing Gustavus' intentions, Ferdinand II sent Tilly instructions, insisting he protect his Austrian lands. It seemed the city might be spared. Disastrously for Magdeburg though, the Treaty of Cherasco on 26 April, between France and the empire, restored peace in Italy. This freed Imperial troops beyond the Alps for service in the Holy Roman Empire, including 20,000 men from Wallenstein's army. An enthu-siastic Pappenheim wrote, 'This summer we can sweep our enemies before us. God give us Grace thereto.'

After a conference with Tilly, it was decided to conclude the siege quickly. Before the Protestants could act. Just two days after the treaty, Pappenheim attacked the fortifications on the islands to the north-east around the *Zollschanze* to isolate it. Using boats and the causeways, the islands were successfully stormed. By the following day the entire outer works had been captured, and for the defenders things were slowly beginning to look bleak.

At the Conference of Leipzig, Gustavus implored the Electors of Brandenburg and Saxony to join him in the relief of the city, for without their military alliance he dared not move. The Swedes, camped in Brandenburg, were ironically getting little aid from the very princes he had come to protect. Without General Georg Hans von Arnim, ironically a former Wallenstein officer, and the 16,000–40,000 Saxon Army it would be near impossible to relieve Magdeburg effectively. In fact, by the end of April Gustavus had sent a message to Falkenberg, telling him he must hold out for another two months. The city was well provisioned, so such a timetable seemed reasonable.

By May 1631 the besiegers were within talking distance of the defenders. Significantly, on 4 May, Pappenheim seized the razed northern suburb of Neustadt, on the left bank of the Elbe west of the captured *Zollschanze*, and began the construction of siege batteries. Tilly summoned the city authorities to surrender. The following day saw the defenders' reply in the form of three brave sorties launched against the siege lines.

Then, on 10 May the town council sent Tilly a message saying that they wanted the mediation of the Electors of Brandenburg and Saxony. Tilly's answer, two days later, was to insist first on surrender. Falkenberg, in the meantime, had sent a desperate but futile plea for help to Gustavus and the Swedish Army.

Early in May Gustavus tried to force his brother-in-law, George William, the Elector of Brandenburg to join him, demanding the occupation of the fortresses of Küstrin and Spandau. On 13 May, Gustavus appeared before Berlin

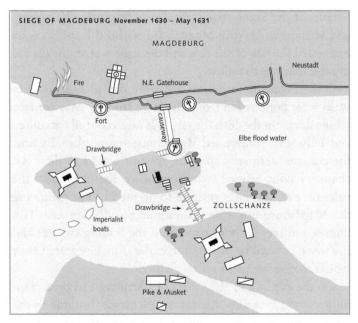

Storming of the *Zollschanze*, 28 April 1631.

to hold a conference with George William. The king warned bitterly:

> I am marching upon Magdeburg, to deliver the city. If no one will assist me, I will retreat at once … But you Protestants will have to answer at the day of judgement that you would do nothing … Magdeburg will be taken, and, if I retire, you will have to look to yourselves.

Brandenburg finally allowed the occupation of Spandau on 15 May. In contrast, John George of Saxony was in communication with the emperor, and refused to deal with the Swedish king. Gustavus still dared not move towards the Elbe because of the Saxon Army. Finally, in frustration he gave up and withdrew, leaving Magdeburg without hope. Saxony had sealed the fate of the city and the stage was set for the sad disaster that was to follow.

Once Pappenheim's batteries were completed, the city walls were bombarded from Neustadt from 17 May for two whole days. On the 18th the city was unsuccessfully assaulted and Tilly again summoned Magdeburg to capitulate. By now the citizens were very anxious, for this would be their last chance; having refused two summonses, the city, under the rules of war, was now entitled to no mercy. The following day Magdeburg was once again unsuccessfully stormed. This intense military activity prompted the leading citizens and the town council to discuss surrender, for they feared they would receive no quarter if taken forcibly.

On the evening of the 19 May Falkenberg had been planning a sortie, possibly against the Neustadt, but due to the lull in Imperialist operations he diplomatically called it off. Falkenberg said he wanted to be consulted on the results of the meeting the next day. The garrison commander had another reason for calling off the sally, he was harbouring an unwelcome secret; the defenders' gunpowder was all but expended. The garrison, perhaps optimistic that the Swedish Army was en route, had used up their ammunition in an almost reckless manner. Falkenberg may have confided in Christian William, but few can have appreciated the situation the city was now in.

Outside the walls the Imperialists were also beginning to get desperate for a swift resolution, fearing the arrival of the Swedish Army. If they failed to take the city soon they could be trapped. To the north was Wallenstein's unhelpful Mecklenburg, south was Arnim with the Saxons and to the east was Gustavus. Indeed, Wallenstein was so resentful of

Tilly he steadfastly refused to send any supplies. Pappenheim and his master were counting on getting their hands on Magdeburg's food supplies and magazines, little realising that there was no gunpowder to be had. Their soldiers were increasingly tiring of the siege and the city's defiance.

At 5 a.m. on 20 May, part of the garrison was stood down and Falkenberg went to the *Rathaus* to talk with the town council. About 30 minutes later, word reached him that Imperialist troops were moving towards the walls and that soldiers were massing in Neustadt. Pappenheim, tired of Tilly's cautiousness, had decided to lead his men in a final assault without orders. At about 7 a.m. he led the attack from Neustadt across the Elbe, with his foot soldiers and Croat troops. Despite the warning, the defenders were caught unprepared and Pappenheim's men swarmed over the river and reached the city wall. Monro records the walls were 'black with men and scaling ladders'. Falkenberg meanwhile, gathering all the men he could, rushed to the northern wall.

To the south, Tilly's adjutant general, Duke Adolphus of Holstein-Gottorp, forced possibly by Pappenheim's actions, led the Imperialists in an assault on the southern walls. The city was struck in a classic pincer attack. Nevertheless, Holstein-Gottorp's troops were held off and Falkenberg's counter-attack caught Pappenheim's men in possession of the northern wall. The Imperialists were thrown back and their attack stalled around the walls.

Pappenheim blamed Tilly for not supporting him, which may be true, but numbers were eventually to tell. A gate was fired and then forced, and several houses were set on fire to hamper the defenders. The Croats secured a second gate and Imperialist forces streamed into the city. During the fighting that followed, Falkenberg was mortally wounded, carrying out his orders to the last.

The struggle raged all morning with the defenders contesting the streets. Inevitably Pappenheim forced his way to the southern walls, taking the defenders in the rear and Holstein-Gottorp's troops also entered the city. By 1 p.m. Magdeburg was completely in Imperialist hands, but its agony was only just beginning. Because the defenders had refused to surrender as anticipated they could expect no quarter. The victors embarked on a killing spree of the garrison and citizens alike. Pappenheim, only by force, managed to save the wounded administrator, Christian William.

Discipline completely broke down, but the rape and pillage were eventually interrupted by fire. In the late afternoon flames appeared in numerous places, possibly due to the excesses, Pappenheim's earlier fires or embittered inhabitants. It was even rumoured that Falkenberg had planned the blaze to deny the Imperialists the city. Certainly Pappenheim and Tilly would not wish to destroy the very place that could provision and pay their bedraggled army. Responsibility has never been proved, although the debauchery taking place on the streets was the most likely culprit.

Some of the soldiers tried to fight the conflagration, for example at the *Liebfrauenkloster*. Tilly himself saved a baby from its dead mother's arms and ordered the prior to herd the women and children into the sanctuary of the cathedral. Some 600 people found safety there, but in the nightmare that followed the city was lost, with upwards of 25,000 of its inhabitants.

For three days Magdeburg burned, until eventually amongst the desolation of the wooden houses only the cathedral remained. For fear of plague, the charred corpses were consigned to the Elbe. It took fourteen days to clear the bodies. Pappenheim, surveying the devastation, was of the view that no such visitation of God had been seen since the destruction of Jerusalem. The rape of Magdeburg was complete, and a shudder ran through the Protestant principalities of Germany. The cry of 'Magdeburg quarter!' would become a common refrain upon vengeful Protestant lips.

All the remaining food stocks had gone up in smoke, but the ravaging soldiers found wine casks intact in the cellars and for two days they drowned out the horror. Most of the female survivors were forcibly incorporated into the Imperialist Army as camp followers and prostitutes. Tilly reasserted some control on 22 May, and three days later a cannon announced the city's return to the true faith. Magdeburg was renamed Marienburg, after Tilly's patroness. Tilly wrote to Maximilian of Bavaria with evident regret, 'Our danger has no end, for the Protestant Estates will without doubt be only strengthened in their hatred by this.'

The appalling loss of the city denied Tilly the most strategic point on the Elbe and its provisions. Nor was there time for celebration or reflection as he was uncertain as to what Gustavus' next move would be. Pappenheim pressed for pursuit, although the remaining magazines of Magdeburg were lost to them and Wallenstein remained recalcitrant.

Tilly did not move from Magdeburg until the end of May with about 25,000 men, while Pappenheim stayed on the Elbe. Instead of marching north-east towards Gustavus, Tilly retired south-west to stop the Landgrave of Hesse-Cassel raising troops for the Protestant cause. He was to fail, for the Landgrave had 17,000 men under arms by November. In fact, Pappenheim recalled Tilly, and by late July was at Wolmirstedt, below ruined Magdeburg.

News of Magdeburg's brutal sacking can only have caused dismay amongst the Protestant ranks. It had been sacrificed to their needless indecision and served as a dire warning to those who opposed Imperial rule. Despite the outrage over the

loss of Magdeburg, Brandenburg and Saxony were still not
stung into action. Nevertheless, regardless of their inactivity,
Ferdinand II would not tolerate the threat they represented.
In June, now that it was too late, Gustavus Adolphus moved
on Berlin, forcing Brandenburg's hand. Also, Greifswald
finally fell to Horn and Gustavus, leaving part of his forces
on the Oder, moved towards the Elbe. This was the start of
a campaign that would take him into the heart of Bavaria.
After capturing Havelberg on 22 July 1631, his army built a
well-fortified camp at Werben in a confluence of the Elbe and
Havel. He remained too weak to take the offensive however,
with an army of probably no more than 16,000.

In early August, Tilly arrived with at least 25,000 soldiers to give battle. His advance under fire from the town
and nearby earthworks became disordered and was driven
off. Several days later, a second attack failed. Finally, having
lost 6,000 casualties and deserters, he realised it was futile to
assault such a strongly entrenched position and withdrew.

The Swedes' defensive action at Werben served its purpose
by weakening the Imperial Army. Nonetheless, Tilly retired
south of ruined Magdeburg and collected troops of both the
Catholic League and empire and by the end of August had
about 40,000 men under arms. It was now that he received
orders from the emperor to force John George to lay down
his arms, because in Imperialist hands Saxony would protect
neighbouring Habsburg's lands and prevent Bohemia and
Moravia becoming the front line.

Tilly crossed the Saxon border while Pappenheim seized
Merseburg and forced Leipzig to surrender on 14 September.
John George was incensed and only now did the Saxon Elector
finally agree to an alliance with Gustavus. Conflict between
the German Catholic and Protestant forces could no longer
be delayed. The scene was set for the bloody engagements

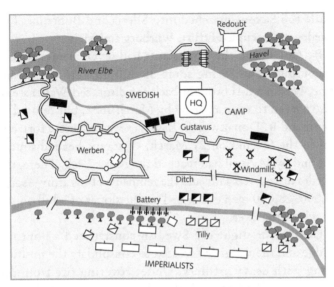

Assault on Werben, August 1631.

of Breitenfeld, Rain (Lech Crossing), Alte Veste (Rednitz), Lützen and finally Nordlingen. The Swedes signed a treaty with the Saxons and united with them at Düben. They confronted the Imperialists forces at Breitenfeld on Wednesday, 17 September 1631.

Gustavus Adolphus had a combined army of about 42,000 against 35,000. The battle went against the Imperialist forces from the start. Tilly was wounded in the chest, neck and, with his right arm broken, fled. Beneath the descending fog and smoke, Pappenheim led the desperate rearguard action against Swedish pursuit. The Imperialists lost some 12,000 casualties, 7,000 prisoners (most of whom defected), 100 standards and twenty cannon to the Swedes' 2,100 casualties. In the eyes of the Protestant world, Magdeburg had been avenged. The Swedish General Baner occupied the ruins of Magdeburg while the Saxons marched into Silesia and Bohemia. The Swedes also occupied Erfurt, Würburg and Mainz.

By Christmas, the Protestants had seven armies within the empire, numbering some 80,000 men. In the meantime, Ferdinand II brought back the disgraced Wallenstein, who agreed to raise an army by March 1632, but refused to command it. Donauworth, Bavaria's westernmost fortress, fell to the Swedes on 27 March. Gustavus headed southeast and by 14 April had reached the River Lech. There, the Swedes discovered Tilly and the remnants of his army – some 10,000–16,000 men encamped on the slopes of a hill on the far side of the river.

During the night the Swedish constructed a pontoon bridge and the vanguard crossed to consolidate the position along with heavy artillery. Under covering fire from the earthworks, the rest of the Swedish Army streamed across the river and surged up the slopes of the hill. Again, the day was not Tilly's, for he was hit in the leg and carried from the field of battle. His subordinate, Johann von Aldringen, was wounded in the head and with the main commanders out

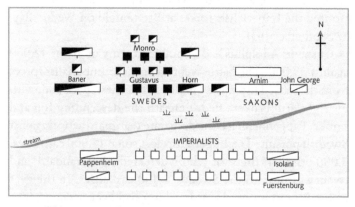

Breitenfeld, 17 September 1631.

of action, Duke Maximilian of Bavaria ordered a retreat. In doing so, he lost most of the army's artillery and baggage. It also left the Swedes free to ravage Bavaria, which they did, destroying some 900 settlements.

Tilly was now altogether out of things with no viable fighting force, and everything rested on the shoulders of Wallenstein. Tilly, at the age of 73, died on 29 April 1632 at Ingolstadt and the emperor finally persuaded Wallenstein to lead his new army. On 25 May 1632 Wallenstein marched into Prague and proceeded to drive the Saxons out of Bohemia while the Swedes were still in Bavaria. He then joined Maximilian of Bavaria, creating a powerful joint force of 48,000 men.

The Imperialists headed for Saxony, drawing the Swedes to them, and the two armies converged west of Nuremberg. In early September the Imperialists retired to a fortified encampment, the 'Alte Veste', on the banks of the River Rednitz, thus threatening Gustavus, who was camped at Fürth. The Swedish, however, had been reinforced and Gustavus felt strong enough to take to the offensive. On 3 and 4 September they assaulted the Imperialist encampment but were repulsed. Gustavus offered peace, but the Imperialists refused to negotiate, so on 18 September the Protestants withdrew northwards, their position and numbers too weak for any further military action.

The Imperialists lost no time taking advantage of the situation, rather than chase the Swedes they decided to eliminate the Saxons. On 9 October, dithering John George despatched a message to Gustavus saying he was under threat from the Imperialist Generals Wallenstein, Holk and Pappenheim, but Gustavus had already realised the situation and was en route. On 1 November Wallenstein took Leipzig and he and Pappenheim united, creating a situation Gustavus must have feared. With no sign of the Saxons joining them, the Swedes numbered only 16,000, in sharp contrast to the Imperialist forces of over 31,000. He found the Imperialists entrenched at Lützen, about 15 miles south-west of Leipzig, but it was too dark to attack and the Imperialists continued to entrench themselves.

The fateful day of 16 November 1632 started off extremely misty. Surveying the field, Wallenstein must have started with deep regret, for Pappenheim with some 8,000 men was at Halle 35 miles away. He had 12,000–15,000 ill-armed foot and 8,000 horse, which still gave the Imperialists a total of 20,000–23,000 men and a superiority of 7,000 over the Protestant Army. If Pappenheim, who had been recalled, arrived in time they would have a crushing total of 31,000.

Despite Pappenheim's appearance, the day saw another Swedish victory, although the Swedes paid a terrible price and their military machine received a body blow from which it would never recover. They crucially lost their warrior king and 10,000 soldiers. The Imperialists also received a terrible blow – the rash Pappenheim, probably the true perpetrator of Magdeburg's sacking, was dead, as well as 12,000 men. The fact that the Swedes suffered fewer casualties and were left in possession of the battlefield only just made them the marginal victors.

After the death of Gustavus Adolphus, the Protestant cause was left in disarray and it seemed it might collapse. Sweden's Chancellor Axel Oxenstierna, with interference from France, set to work to keep the allies together with the League of Heilbronn. The powerful and independent Wallenstein was assassinated at the end of February 1634, a victim of court politics.

Three months later, King Ferdinand of Hungary (Ferdinand II's son) and Count Matthias Gallas, with about

25,000 Imperialist troops moved from Pilsen. Also, Johann von Aldringen with about 8,000 men marched on Nordlingen. This fortress town could provide a strategic base, from where the Imperialists could deploy their troops in Swabia. Swedish General Bernard, realising this, called upon the inhabitants to resist, promising to relieve them as soon as possible, and shortly afterwards Ferdinand arrived with 15,000 Imperial troops.

By 23 August 1634 the redoubtable Bernard and Horn, in the neighbourhood of Nordlingen, were reluctant to commit their tired army with the approach of Ferdinand of Hungary's cousin, Ferdinand Cardinal-Infant of Spain, with at least 15,000 Spanish troops from Milan, en route to the Spanish Netherlands. On 2 September the Spanish arrived and Horn wanted to wait for the Rhinegrave Otto Lewis, who was approaching with 6,000 troops of the Swabian Circle.

The Imperialists outnumbered the Protestants by nearly a third. The Austrian and Spanish Habsburgs could muster 33,000 men: 20,000 foot, including the cream of the Spanish forces, and 13,000 horse. The Protestants, under Bernard and Horn, could muster at the most 25,000 men: 16,000 foot and 9,000 horse. Tragically for Horn, many Swedish troops had been sent to Poland to fight in the War of Smolensk.

On 6 September 1634 Horn's forces were in position and he planned to attack the Imperialists' hill defences with his Swedish foot, followed by a cavalry charge in their far flank. The Swedish horse attacked prematurely, chasing off a few odd Imperialist units but leaving the infantry unsupported. In the struggle that followed Horn was captured and Bernard only just escaped after a Protestant dragoon gave him his nag.

The Protestant Army lost all its artillery, baggage, standards and many senior officers. Casualty figures for the Protestants vary, the highest total recorded being 17,000 killed and 14,000 prisoners. Oxenstierna puts the figure at 12,000 killed and captured. Whatever the Protestant losses, the Imperialists, in sharp contrast, lost a mere 1,200 men.

The remnants of the Swedish forces rallied at Heilbronn with the Rhinegrave's troops. Nordlingen had little option but to surrender. All that had been won at Breitenfeld and Lützen in the wake of Magdeburg was swept away. The Heilbronn League collapsed and south-west Germany was overrun by rampaging Imperialist armies. Ungrateful Saxony and many other estates abandoned the alliance with Sweden and made peace with the emperor.

Swedish predominance in Germany was largely finished. A fresh alliance was concluded in 1635 between Sweden and France at Compiegne, but on 19 May 1635 France declared war on Spain and the Thirty Years War began to move in a new direction.

The Imperialists sent troops to Brandenburg to reinforce the Saxons. However, in September 1636 Baner marched from his camp at Werben and cut their joint forces off at Wittstock on the Dosse, a tributary of the Havel on the Brandenburg–Mecklenburg frontier. The Imperial and Saxon forces under the Elector of Saxony and the Imperialist General Count Hatzfeld were defeated at Wittstock on 4 October 1636. Protestant victories followed at Rheinfelden on 2 and 3 March 1638 and Wittenweier on 9 May 1638.

By 1641 the exhausted Protestant Army was in a dire condition, although it still managed to resist the Imperial forces. The Swedes secured a victory over Archduke Leopold William at Wolfenbüttel on 29 June 1641, but it was not this victory but rather the timely arrival of Lennart Torstensson and 7,000 fresh troops from Sweden that saved the allies from dissolution; he was besieging Leipzig when Leopold's force arrived on 2 November 1642.

The Swedes, with the weaker army of over 10,000, withdrew northwards in the direction of Breitenfeld. Leopold, with over 20,000 men and hoping for a second Nordlingen pursued, but it was to be a second Breitenfeld. The battle opened with the Imperialist artillery covering the horse as it formed up. They were using chain-shot, which was still quite an innovation at this date. Torstensson realised he had to attack the superior Imperialist Army before it was fully deployed and before his own army became too dispirited. The archduke lost about half his army and only just escaped, retiring to Bohemia.

Four years later, on 2 March 1645 at Jankau, about 9 miles from Tabor, a mixed force of Imperialists, under Count Hatzfeld, and Bavarians, under Count von Goetz, Franz von Mercy and Johann von Werth, cut Torstensson's Swedes off and forced him to fight. Both armies numbered in the region of 16,000, although it seems the Imperialists were superior.

The Bavarian horse and Imperialist reserves attempted to hold the Swedes at bay, but with half their army gone, on unfavourable ground, they suffered heavy casualties and were surrounded before being routed. Hatzfeld was captured, along with 4,000–5,000 men, while the remnants of the Bavarian and Imperialist horse fled towards Prague.

Swedish forces under Count Palatinate Charles Gustavus (later Charles X of Sweden) ended the war in 1648, besieging Prague. That year witnessed the Peace of Westphalia, which finally brought an end to the ruinous Thirty Years War and allowed Sweden to disengage from eighteen years of largely fruitless warfare, leaving France and Spain to fight on until 1660.

In hindsight, the tragic destruction of Magdeburg served little purpose and simply denied the Catholic forces the most strategic point on the Elbe. Furthermore, its people were needlessly sacrificed to the indecision and political infighting of the German Protestant princes. The siege did, however, ensure that Tilly and Pappenheim's forces remained tied up during May 1631 while the Swedes strengthened their position. Only its full destruction finally stung Gustavus Adolphus and the Protestant Electors into taking action.

The impetuous Pappenheim takes full responsibility for Magdeburg's devastation. If he had waited even just one more day, the city might well have capitulated; instead, he unleashed his soldiers to do their worst. Once the town council discovered the garrison's gunpowder had been used up, surrender would have been inevitable. Tilly and Pappenheim's pyrrhic victory was short lived, for within a year the pair were dead and the Imperial cause ultimately gained very little from the brutal sacking of the city. In the annals of siege warfare its fate clearly illustrates what any city could expect if it has to be taken by storm.

9

Turning Point: Lostwithiel 1644

Marston Moor undoubtedly dominates the military engagements of the English Civil Wars; the second largest battle ever fought on British soil after Towton in 1461. In strategic terms, it was the penultimate battle of the First Civil War of 1642–46.

Naseby, fought in 1645, was the final death blow to the Royalist cause of King Charles I. Strikingly, though, shortly after Marston Moor, south-western England was the scene of a unique Royalist victory which proved just as significant. The Battle of Lostwithiel was the largest Royalist success in the whole war, indeed south-western England not only witnessed the opening shots of the civil war and Charles I's finest moment at Lostwithiel, but also the Royalists' final death throes at Torrington.

Opposition to Charles I's reign arose from his centralisation of authority in order to raise money. In doing so he fundamentally undermined and divided county support that would be fundamental to his survival during the 1640s and the First Civil War. Inexorably, the first fifteen years of his rule confirmed the alienation of the country from the court.

The road to war was marked by the Petition of Rights, the 1629 Dissolution, Hampden's case and the Bishops' Wars.

Charles' financial problems stemmed largely from an over-aggressive foreign policy and continuing antagonism with Parliament. Despite his dangerous lack of royal funds, he persisted in pursuing an expensive foreign policy and by 1627 England was at war with both France and Spain. Two years later Charles I was still badly in need of money and Parliament adopted an aggressive stance. In return for funding, he was presented with 'the Petition of Rights', stating that Parliament wanted: no billeting of troops on private quarters; no arbitrary imprisonment; no commissions of martial law and no forced loans. Instead of agreeing, Charles decided to dissolve Parliament and rule alone.

It was during the period known as the 'Eleven Years' Tyranny' of 1629–40 that his extraordinary means of finance really began to stir up opposition. Without Parliament to collect taxes, Charles was forced to raise money via royal prerogative and feudal dues. In 1631 he issued his Book of Orders, heavily influenced by his chief ministers, William Laud, Archbishop of Canterbury, and Thomas Wentworth, Earl of Strafford, especially their policy of 'Thorough'.

This was seen as an infringement on the rights and privileges of the people and ultimately contributed to the breakdown of Charles' government in 1640. It was 'Thorough' that particularly caused the king unnecessary unpopularity and Laud's implementation of Charles' religious policy in Scotland resulted in the Bishops' Wars of 1639 and 1640. While the levying of Ship Money was accepted as a reasonable necessity, when the king levied it on the whole country it was seen as another arbitrary way to raise extraordinary revenue. John Hampden of Buckinghamshire refused to pay in 1637 and was brought before a court. He lost and duly paid up the sum for which he had been assessed, but the king won a pyrrhic victory.

With the hostility increasing between the king and Parliament, William Seymour, Marquis of Hertford, was appointed lieutenant general of the south-west on 2 August 1642. Leaving the king at Beverly, Hertford moved to Bath and then into Somerset. The local Parliamentarians (or Roundheads) gathered their forces and the Royalists retired to Somerton, successfully defeating them at Marshall's Elm two days after his appointment. William Russell, Earl of Bedford, was then despatched by Parliament to deal with Hertford's Royalist forces.

On 22 August 1642, King Charles I raised the royal standard at Nottingham, and war was officially declared. Then, on 23 October, the king's forces and those of Parliament, under the Earl of Essex, clashed in the Midlands at Edgehill. The outcome was indecisive, but there was no going back. The end of the year saw a further clash in the north, at Tadcaster on 6 December.

Royalist Sir Ralph Hopton marched into Cornwall, raising the local trained bands, and invaded Devon, although his attempts to blockade Plymouth and seize Exeter failed. With the approach of a Roundhead force under General Ruthin he retired into Cornwall and turned to give battle at Braddock Down on 19 January 1643. Securing victory, the Royalists moved back into Devon – one column drove Ruthin out of Saltash, while a second pursued Henry Grey, Earl of Stamford, the new Parliamentarian commander-in-chief of their south-western forces.

Hopton confronted and defeated Stamford at Stratton on 16 May 1643. Leaving forces to blockade Exeter and Plymouth, Hopton rendezvoused at Chard with the king's nephew, Prince Maurice, and Hertford. Parliamentarian

General Sir William Waller, having secured Gloucester and Bristol, moved to Bath. On 5 July 1643 the two armies gave battle at Lansdown. Waller goaded the Royalists into attacking and Hopton stormed up the slopes of the hill in the face of the Roundheads' muskets and artillery. Waller launched a counter-attack with his cavalry, but could not dislodge the Royalists from the edge of the hill. Both sides fought to a standstill, but with darkness Waller withdrew.

The two sides clashed again on Roundway Down on 13 July 1643. The Royalists horse routed both wings of Waller's cavalry, and Hopton's infantry marched out of Devizes to help crush the Roundhead foot. The whole of Waller's army was shattered.

With Sir William Waller's forces temporarily knocked out of the war, the king was presented with an ideal chance to finally seize Bristol, the second city of the realm. On 18 July, Prince Rupert, the king's German nephew, left Oxford with an army to join up with the victorious Royalist western forces. The combined armies of Prince Rupert and Prince Maurice (Rupert's brother) numbered about 20,000 men, against which Bristol's Governor Nathaniel Fiennes had about 2,000. The outcome was inevitable.

Royalist fortunes were at their height in 1643, but the following year was to see dramatic events that would spell the beginning of the end. First, influence in the south was lost with Hopton's defeat at Cheriton on 29 March 1644 by his sparring partner Waller. Even so, the king's cause was reasonably healthy in the south-west, the Midlands and the north.

The new danger that emerged was the Scots' entry into the war on Parliament's side. The king, at Oxford, was threatened by two Parliamentarian armies while the Royalist forces in the north were menaced by three. The whole balance of power in the north was threatened with collapse, and the Royalist Earl of Newcastle was forced on the defensive at York.

The Scots Covenant Army, under the Earl of Leven, and the Parliamentarian Northern Army, under Lord Fairfax, laid siege to York on 22 April 1644, one of the north's principle cities and a Royalist bastion, trapping the Earl of Newcastle and his small force. Prince Rupert departed Oxford on 5 May for the relief of the city, but the following day the Parliamentarian Earl of Manchester took Lincoln and joined Fairfax and the Earl of Leven at the siege of York. On 19 May, the Earl of Essex and Waller agreed to act as separate commands against King Charles' Oxford Army, which had absorbed the remnants of Hopton's troops following Cheriton.

Prince Rupert left Shrewsbury on 16 May, though his army (some 6,000 foot and 2,000 horse) was insufficient to confront the combined forces of Leven and Fairfax. Sensibly, Rupert decided to march via Lancashire to collect reinforcements and to mop up Parliamentarian garrisons. He also planned to link up with Sir Charles Lucas, who had escaped from York with about 2,000 cavalry. On 25 May, the prince seized Stockport, followed two days later by Bolton. Lord George Goring and Lucas reinforced Rupert on the 30th with 5,000 horse and 800 foot. The Royalist Army continued its steady march north, and Liverpool fell to it on 11 June.

Meanwhile, on 6 June King Charles' army arrived at Worcester having successfully evaded Waller and Essex. The latter went to relieve Lyme, which was under siege, leaving Waller to watch the king. Charles, on hearing this, gained reinforcements from Oxford and prepared to deal with Waller. On 29 June 1644 Waller attacked the Royalists across the River Cherwell, crossing at Cropredy

Bridge. His army numbered some 5,000 horse and 4,000 foot against the king's 5,000 horse and 3,500 foot. Attacking the divided Royalist Army, Waller was cut off by the Earl of Cleveland's horse and he had to fight his way back, losing his artillery and 700 men, to the loss of about 100 Royalists. A bruised Waller withdrew.

On 7 July Charles and his council of war decided to move south-west in pursuit of Essex. In the meantime, the Royalist Sir Richard Grenville was forced to abandon his siege of Plymouth. Parliamentarian Lord Robartes persuaded Essex that the Cornish would rise up for Parliament, and they could stop the Royalists exporting Cornish tin. As a result, Essex's army invaded Cornwall.

King Charles, at Tickenhill, was anxious to have events in the north settled. His Oxford Army at present had to keep the armies of Essex and Waller occupied, with little hope of substantial aid. Under the guidance of Lords Wilmot and Digby, Charles wrote a fateful letter to Prince Rupert on 14 July. It began with the famous line, 'If York be lost I shall esteem my crown little less', and went on in unclear terms to outline his desire for Rupert to relieve York and defeat Leven and Fairfax. The letter was extremely ambiguous, but to the prince his honour could dictate only one meaning – he must fight. When Lord Culpeper heard of the letter, he exclaimed to the king, 'He will fight, whatever comes on't'. Disaster was potentially looming for the Royalist cause and the dice had been cast.

In York, Newcastle's trapped garrison mustered a little over 4,500 men, while outside the city walls Fairfax had an army of about 5,000 and Leven's Scots numbered 16,000. To make matters worse, on 3 June the Earl of Manchester arrived with Parliament's Eastern Association Army of 9,000, bringing the combined Allied contingent up to about 30,000 men

York was completely cut off and the besiegers prepared to beat or starve the garrison out.

The fast-moving Rupert had reached Skipton Castle by the 26th and four days later was only 14 miles west of York at Knaresborough. The Allies had hoped to be able to maintain their siege and confront Rupert, but their sources misled them into believing the prince's army was stronger than it really was and they abandoned their siege lines. The Parliamentarian forces moved north-west in preparation for battle deployment and on 1 July a body of Royalist horse appeared before them and they were convinced it was Rupert's vanguard. The three Allied armies converged on Marston Moor, 6 miles west of York. Rupert, though, had deceived his opponents, for his army moved first north and then south to relieve York.

Having been duped, the Parliamentarians were uncertain how to react, but Leven, who was the senior officer, proposed they move south to block Rupert's lines of communication. This they did on 2 July, the three armies marching towards Tadcaster in order to cut Rupert off and to await reinforcements. The prince, in the meantime, was trying to persuade Newcastle to join in with an attack, but the earl was reluctant. Newcastle, like his enemies, wished to await the arrival of reinforcements, also his troops were in a semi-mutinous state over pay.

Rupert marched his army towards Marston Moor at 4 a.m. on the 2nd, Newcastle had not arrived until 9 a.m. and then without any of his foot. The prince's chance of attacking the stretched-out marching allied armies was almost gone. In order to press his point, Rupert informed Newcastle that he had a letter from the king ordering him to attack. The earl conceded to the prince's wishes, but it was too late – only a set-piece battle could now be fought. Leven, on being informed that the Royalists were drawing up on the northern edge of Marston Moor, decided to turn and fight at about 10 a.m.

The deployed Royalist Army was too weak to take the offensive. Newcastle's foot, who had been sent for, did not arrive until about 3 p.m. In all, until the arrival of Newcastle's contingent, Rupert had about 6,800 foot, 7,000 horse and sixteen cannon. The total strength of the allies is uncertain and ranges from 21,000–28,000 plus twenty-five pieces of artillery. Whatever the true figure is, the allies had a potentially decisive superiority of at least 5,000 men.

By about 9.30 p.m. it was all over, Rupert's army had been shattered. Casualties are rather vague, but it would certainly be fair to say the Royalists suffered 50 per cent more than the Parliamentarians, particularly during the pursuit. It is thought that the Royalists lost perhaps 3,000 killed and the Parliamentarians 1,150, but the overall losses including wounded was probably at least 6,000.

The Royalists also lost 10,000 arms and 1,500 prisoners. Newcastle's fighting spirit was broken and he sailed for Holland. Rupert managed to collect 6,000 horse and withdrew to Lancashire, while York – the cause of the entire campaign – surrendered to the Parliamentarians on 16 July.

Marston Moor was not as decisive as it seems. Had the Royalists made an effort, they could have continued the war in the north. The psychological impact was such that the Royalists' will to defend the north collapsed and Marston Moor heralded the beginning of the end for the Royalist cause.

To the south, the king's army marched westwards over the Cotswolds, through Cirencester and arrived at Bath on 15 July. While there, he received news of Rupert's disastrous defeat. It was decided to carry on after Essex, the king wishing to do so because he was anxious for the safety of the queen, who was believed to be at Exeter, but had in fact sailed from Falmouth on 14 July for France.

Essex was under the misapprehension that Waller was on the heels of the king, and he did not realise to what extent Waller had been mauled at Cropredy Bridge. The earl headed for Weymouth, then Crewkerne and Chard; arriving at Exeter late in June, he found the queen gone. Essex then took Taunton early in July, although after leaving garrisons in Dorset and Devon his army was beginning to diminish. On 23 July, he arrived at Tavistock and then sent a messenger to the Committee of Both Kingdoms (England and Scotland) in London, informing them of his intentions to relieve Plymouth and hoping Waller was keeping the king occupied.

Royalist Sir Richard Grenville, with his lines of communication threatened, raised his ineffective siege of Plymouth. He retired across the Tamar to Saltash, collecting its garrison as well as Mount Stamford and Plympton's. Grenville then proceeded north to guard the bridge over the Tamar at Horsebridge.

Having achieved his intention, Essex was now faced with the decision of whether to turn and fight the advancing Royalists, or to carry on into Cornwall. The prize of Falmouth, Lord Robartes' promises that the Cornish would rise up against the king, and the stopping of the export of Cornish tin which was paying for Royalist munitions imports, finally persuaded him.

Convinced that Waller was pursuing the king, Essex decided to march west supported by the navy under the Earl of Warwick. He strengthened his army with a regiment of foot and horse from Plymouth's garrison, much to Warwick's consternation. On 26 July, Essex invaded Cornwall, sending Sir Philip Stapleton back to London to explain his actions to the House of Commons. The Parliamentarian advance guard overcame and drove off Grenville's troops at Horsebridge. On the same day, Charles reached Exeter and united with Prince Maurice. King Charles I now had 16,000 men and if Grenville joined him as expected he would have some 20,000.

Essex only had 10,000 men under arms, but he continued through Liskeard to Bodmin, Grenville falling back before him to Truro. On 2 August he heard the king was at Launceston, 20 miles away to the north-east. Abandoning any ideas of taking Falmouth or Truro, he decided to make contact with Warwick's fleet at Fowey. Withdrawing southwards to Lostwithiel, he left a small force at Bodmin and despatched others to take the port of Fowey. However, unbeknown to Essex, a Royalist trap was slowly forming. On 3 August, King Charles ordered Grenville to move to Tregony, 15 miles south-west of Lostwithiel.

Now aware of the potential danger, the following day Essex informed the Committee of Both Kingdoms that he was under threat from three Royalist armies – those of Charles, Maurice and Hopton – also, he was gaining no support from the local population. On arrival at Fowey, Essex found no sign of Warwick or his ships. Furthermore, when Essex heard that Charles was at Launceston, he was in fact really at Liskeard and on the 4th the Royalist vanguard captured Lord Mohun's house, Boconnoc, some 3 miles east of Lostwithiel.

Essex was becoming increasingly alarmed, for he needed a large forage area for his troops and Fowey for supplies or possible evacuation. He decided to base his defence on Lostwithiel and its bridge, the lowest crossing point of the River Fowey. His right wing and communications to Fowey were protected by the river's deep gorge. Essex sent 1,000 foot to secure Fowey itself, leaving himself a front-line army of 7,000 foot and 2,000 horse. The basis of his defence line was the old shell keep of Restormel Castle, mounted on a hill a mile north of Lostwithiel; another hill 1 mile north-east of Lostwithiel on the right bank of the Fowey; Beacon Hill, 1 mile eastwards, and Druids Hill, about 1½ miles to the north-east. At each of these points, Essex embattled 1,000 foot, except for Druids Hill where there was only an outpost, with another 1,000 to the west of Restormel and 2,000 between the hill north-east of Lostwithiel and Beacon Hill. The bulk of his horse were at Trewether to the east and Chark to the west. In the meantime, Waller was still not ready, so he despatched Lieutenant General Middleton with 2,000 horse to aid the hapless Essex.

On 11 August, Grenville arrived at Bodmin with 2,400 men, driving out the Parliamentarian horse. On the 12th, he took Respryn Bridge over the Fowey and linked up with Charles on the left bank. The following day, Lord Goring and Sir Jacob Astley captured Polruan Castle, sealing off Fowey Harbour and the east bank of the river. Disastrously, on the 14th Middleton's horse were defeated at Bridgwater by Sir Francis Doddington and they retreated to Sherborne. For a week the Royalists made no move, but Charles realised Essex must be defeated before further relief arrived. Warwick, who was stuck at Plymouth, wrote to Parliament on the 18th, informing them that the winds were too westerly and he could not reach Essex's army.

Royalist forces both west and east of Fowey made a co-ordinated advance on 21 August 1644. At 7 a.m., marching out of the morning mist, Grenville's troops managed to seize Restormel Castle – its garrison, John Weare's Devonshire Regiment of Foot, retired without orders. On the right, Maurice's foot took Druids Hill and the Earl of Brentford seized Beacon Hill, where surprisingly opposition was minimal. Still under the cover of the mist, Maurice then

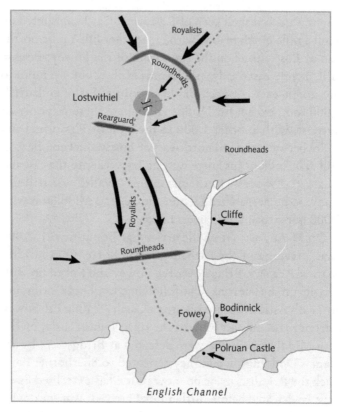

Lostwithiel Campaign, 14–31 August 1644.

sent a column of 1,000 men to take the hill north-east of Lostwithiel. By nightfall the Royalists had taken all the high ground north of the town.

The Earl of Essex's hopes were sagging, there was no sign of Waller's army or Warwick's fleet. Also, his troops were running low on ammunition and rations. Instead of doing anything bold, he decided the horse should break out and the foot should retire to Fowey, in the hope of being picked up by Warwick. The night chosen was the 30/31 August, thus taking advantage of the Royalists' 15-mile dispersal. Sir William Balfour was ordered to take the Parliamentarian cavalry up the main Lostwithiel–Liskeard road.

Late on the 30th, two deserters arrived at Boconnoc House and revealed the whole plan, except for the direction of the break-out. All the Royalist posts were alerted and told to stand to arms. The Lostwithiel–Liskeard road was guarded by a cottage with fifty musketeers.

Unlike the rest of Essex's lacklustre campaign, the escape of the Parliamentarian horse was not ill fated. At about 3 a.m. on 31 August, in the dark and mist, Balfour left the town. The musketeers in the cottage were asleep and when they did realise what was happening it was too late. At dawn, Cleveland gave chase with 500 horse, but at Saltash he was held off while Balfour's troopers were ferried across the Tamar.

The Parliamentarian horse successfully reached Plymouth with the loss of only 100 men. Essex's infantry now retreated, marching south from Lostwithiel through the village of Castle down to Fowey. Major General Sir Philip Skippon was in command of the rearguard and he was ordered to hold the high ground north of Castle.

At 7 a.m., 1,000 Royalists under Prince Maurice forced their way into Lostwithiel, driving off Skippon's men who were trying to destroy the Fowey Bridge. King Charles then ordered up two guns, to be placed in Essex's old camp in order to bombard Skippon. Charles, leading his Life Guard of Horse, rode from Beacon Hill at 8 a.m. and forded the Fowey just south of Lostwithiel. Also about this time, Grenville's foot began to advance, skirting the town. Skippon, fearing encirclement, retreated and the Royalist foot pursued. The king also gave chase with his Life Guards. He pursued for

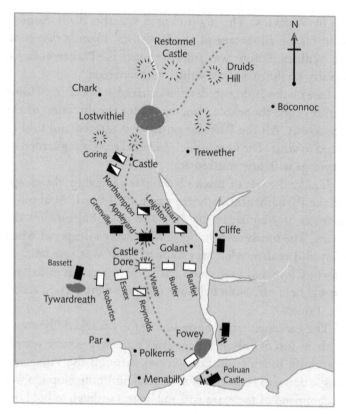

Castle Dore, 31 August 1644.

2 miles, pushing the Roundheads back between Tywardreath and Golant.

Although Grenville's foot arrived to join the fray, they were driven back at about 11.30 a.m., then rallied on the King's Life Guard of Foot under Lieutenant Colonel William Leighton. The Life Guard of Horse then charged, with Major Edward Brett leading the Queen's Troop. They drove back Skippon's soldiers defending a hedgerow. Brett was wounded in the left arm by a musket ball and after the action Charles knighted him.

The Parliamentarian line of defence now ran from the high ground east of Tywardreath to the heights west of Golant, their centre being at Castle Dore. A battery of guns was sited at Castle Dore, and to the west were Essex's and Robartes' regiments, while to the east were the regiments of Colonels Weare, Butler and Bartlet. The valiant Skippon, still pressed by the King's Life Guards, retired to Essex's new defensive line.

By 2 p.m. the bulk of the Royalist foot had arrived. Colonel Basset came up from St Blazey with his 1,500 men and attacked Essex's left wing in the area of Robartes, while Colonel Appleyard, in the vanguard of the King's Foot, attacked the centre, pressing forward. The line, however, did not collapse. About 4 p.m. Essex ordered a counter-attack in order to slow down the Royalist advance. Captain Reynolds and the Plymouth Horse, who had remained behind, some 200 men supported by 100 musketeers from Essex's own regiment, were assigned the task. They swept forward and with a crash drove the Royalist foot back several fields. The rest of Essex's regiment under Lieutenant Colonel John Boteler also advanced and the fight went well, with a wedge driven into the Royalist line. Suddenly though, the Life Guards began to move forward, and as a result Reynolds and Boteler had little choice but to withdraw. About this time Goring arrived, with 2,000 cavalry, but foolishly they were sent in pursuit of Balfour.

Two hours later, Essex attempted to retake the high ground north of Castle Dore, which was topped by an old Iron Age fort. Colonel Weare's regiment was in the vanguard and the counter-attack started off well. Once again, the tired Royalist foot were driven back several fields, but the Earl of Northampton's brigade of horse counter-charged the leading Parliamentarian regiments. After a fierce struggle lasting an hour, Essex's forces were driven back to Castle Dore. Fortunately for Essex, the light was failing fast and the Royalists pressed no further. Meanwhile, Weare's regiment, which had borne the brunt of the fighting, suddenly broke and fled southwards. In the gathering dusk Butler and Bartlet, with their left flank exposed, had little option but to retire. The Parliamentarian centre had collapsed and the road to Fowey was open.

Skippon was in a desperate situation, he sent a messenger to Essex for orders, but all the earl would offer was advice, it seems he had lost all heart in the fight. The Parliamentarian Army was completely trapped and the end was in sight. Essex suggested Skippon should hold a line across the Gribbin Head to Polkerris, supplying a point for evacuation. Failing this, he should draw the baggage train up around the foot and threaten to blow up his powder, thus forcing the Royalists to offer reasonable terms of surrender. The Earl of Essex and two other gentlemen boarded a fishing boat and using the westerly winds sailed for Plymouth, abandoning his army to its fate.

On 1 September, Skippon called a council of war. He suggested to his commanders that their men should try to fight their way out, but they informed him the soldiers were too exhausted and hungry, so he decided to surrender. The following day Skippon formally capitulated, his remaining 6,000 men laid down their arms, handing over about 2,000 pikes, 4,000 muskets, 6,000 swords and 100 barrels of powder with match and bullet. The general of the ordnance, Richard Deane, surrendered some forty-nine brass cannon and a large mortar called 'the Great Basilico of Dover'.

The terms of surrender were:

1. The Roundheads should not take up arms against the king until they reached Southampton or Portsmouth.
2. All regiments may keep their colours.
3. All officers and NCO's may keep their swords and personal possessions.
4. A Royalist escort will be provided as far as Poole or Southampton to protect them from plundering.
5. All Roundhead wounded may stay in Fowey until recovered, and then will be provided with passes to Plymouth.
6. Each regiment may take one wagon to carry their baggage in.
7. All weapons, except officers, must be surrendered.

On the afternoon of 2 September the Roundheads, under Royalist escort, began their retreat. Richard Symonds, in the King's Life Guard, noted that the Parliamentarians marched out of Fowey 'pressed all of a heap like sheep, though not so innocent, so dirty and so dejected as was rare to see'. Reaching the streets of Lostwithiel, the men were attacked by angry locals who had scores to settle and property to recover. Skippon rode up to the king and complained bitterly about some Royalist soldiers having robbed him of his pistols, sword and scarlet coat. Charles asked him to point out the offenders. Seven looters were identified and hanged from the nearest tree as an example to the rest of the army. It was sad that Skippon, knowing the problems faced by the common soldier, could not have asked for their pardon.

Reaching Somerset, the Roundheads were handed over to Middleton's horse. Skippon very civilly thanked the Royalist escort, saying 'that they had carried themselves with great civility towards them and fully complied with their obligations'. The Roundheads continued on their journey and by the time they reached Portsmouth their numbers were below 4,000. This was due to desertion, exposure, disease and hostile locals – only 100 joined the king's cause.

As the Parliamentarians left, Grenville, with the Cornish foot and horse, was sent to join Goring in his pursuit of Balfour, who was now sheltering in Plymouth. Instead of directly joining Goring, Grenville reoccupied Saltash on 3 September, seizing eleven cannon plus arms and ammunition. This was a sensible move because the town was needed for blockading Plymouth, but it denied Goring infantry support. Balfour left Plymouth unopposed, though Goring did send a commanded body of horse to follow the Roundhead column. Unfortunately, the Royalists pursued too closely and the Roundhead rearguard fell on them. After a sharp struggle in which Captain Samuel Yeoman was killed, the Royalists withdrew and Balfour rode on to Dorset unopposed.

Goring and his horse, moving eastwards, were summoned to join the king at Oxford. The king, though, fearing Sir Thomas Fairfax and the New Model Army would relieve Taunton, sent Goring and his 3,000 horse back to the southwest. It was a fatal move for the king's cause.

Charles moved north, and he finally gave battle with Fairfax on 14 June 1645 at Naseby. His army only numbered 9,000 against the Roundheads' 14,000 and the Royalists were severely beaten. Had Goring been there, they may have won. Charles retreated to south Wales and Rupert moved back to Bristol. His last remaining horse were defeated at Rowton Heath on 24 September 1645. Also, on 13 September, the Earl of Montrose's small Scots Royalist Army was defeated at Philiphaugh in Scotland.

The king's cause was rapidly dying. Charles returned despondently to Newark. On 16 February 1646, the 10,000-strong Parliamentarian Army made a surprise night attack on

Torrington, scattering Hopton's last viable command. The remaining Royalist field force was defeated at Stow-on-the-Wold on 21 March 1646. King Charles I surrendered himself to the Scots on 5 May 1646 and eventually he would find himself facing the executioner's block in Whitehall.

The summer campaign of 1644 was a great victory for the Royalist cause and had seen both Waller and Essex's forces put out of action by the king's Oxford Army. Its culmination witnessed the humiliating capitulation of the Earl of Essex's army, plus the capture of 6,000 arms, something that the Royalists were always short of.

The campaign, though, has to be placed in context. As early as 29 March, the Battle of Cheriton saw the loss of Royalist influence in the south. Then, 2 July saw Rupert defeated at Marston Moor and, as a result, the whole of the north was lost to Parliamentarian control. Even so, the Royalist cause still fought on in Scotland, Wales, the lower Midlands and the south-west.

Lostwithiel was a complete disaster for Parliament and a great blow to morale, especially after their success at Marston Moor. Without realising it, the Royalists had conducted a manoeuvre campaign that even Napoleon would have been proud of. Systematically the Parliamentarians had been divided and dealt with piecemeal. It proved the king's cause was still fighting fit despite Rupert's defeat. Furthermore, to the embarrassment of Parliament, it was the only time during the civil war that such a large field army surrendered en masse. Essex led his army to defeat and destroyed his career.

In retrospect, it is easy to criticise Essex's ineptitude, but it must be remembered he was outnumbered by almost two to one. It is interesting to note that when he made his report to the Commons on 7 September 1644, they drew up a letter thanking him for his services and blaming Middleton for his defeat. Essex himself admitted, 'it is the greatest blow we have ever suffered'. Lord Robartes, though, must take his fair share of the blame, for it was his misguided counsel which convinced Essex of the merits of moving further west. Even so, the earl should have known better and fought on the ground of his own choosing in Devon or Somerset. To compound his error of judgement, the weather was against his evacuation plan and Parliament's seaborne supplies, but this does not excuse him for failing to secure his western and eastern flanks.

The Restormel line should have held, but the failure to do so was due to Weare, as at Castle Dore. Once in London, Essex placed charges against Weare for his actions and subsequent desertion to the Royalists. He also charged Butler and Bartlet for failing to hold their position east of Castle Dore.

Essex, though, failed to make effective use of his horse. Ideally, they should have beaten up the Royalists' quarters and harassed their rear. Perhaps the captain general's major fault was that he failed to give his troops inspired leadership. He has been summed up as a good-hearted, muddle-headed sort of man, not lacking in dignity or a sense of duty, but devoid of any real grasp of strategy. What other modern general has fought all his major battles with the enemy between him and his base? To many, his greatest failing was his abandonment of his army, a logical but not a very commendable move. To most, he ingloriously escaped. Only his shortcomings and the brooding feeling of betrayal really explain the apathy Essex displayed throughout the campaign. He lacked the skill he had portrayed at the relief of Gloucester on 8 September 1643, and at the First Battle of Newbury on 20 September 1643. In contrast, we can only commend the conduct of Balfour and Skippon, particularly Skippon, for his dogged resistance even when all was lost.

Lostwithiel was the one and only time that the king fought in the south-west. Charles emerges in a completely new light and his leadership proves that the brilliant Lostwithiel campaign of 1644 was a personal triumph for him. He showed himself to be every bit a warrior king, displaying courage, confidence and cunning. Charles proved that he was an able leader of men by his own personal leadership of his Horse Guards and by risking his life in the face of the enemy.

Although the Royalist victory was almost complete, one may wonder why 6,000 Roundhead foot soldiers were allowed to march away. By seventeenth-century standards it would have been impossible and impractical to imprison such a large number of men. Only officers were of any great value for ransom or exchange. The only blot that lies on the Royalists' management of the campaign is the escape of the Parliamentarian horse, but in practical terms there was little they could have effectively done.

Charles' methodical planning gave him a victory, which secured the south-west for his cause and destroyed the future of Parliament's captain general. In short, the campaign of Lostwithiel was a triumph for the King of England and the biggest success obtained by the Royalists in the whole of the war.

10

SWORD V. MUSKET:
FALKIRK 1746

Culloden Moor was the climax of the five Jacobite Rebellions of 1689, 1708, 1715, 1719 and finally 1745. Probably the most serious crisis to face eighteenth-century Britain came to a close with Culloden. However, it was Falkirk that was the real turning point for, despite gaining a notable victory, the Jacobites threw away the opportunity of going back on the offensive in 1746. Instead, they continued north on the road to ultimate defeat. The crisis also spelled the end of the Highland way of life and the clan system.

The rebellion saw the last contested show of arms over the British throne, between the ruling Protestant House of Hanover and the exiled Catholic Stuarts. The Act of Settlement ensured no Catholic monarch could legally sit on the English throne.

The 1745 was not a nationalist affair with Scotland versus England, for all the regular Scots regiments fought for the Hanoverian Government, while a number of Irish and French troops fought for the Jacobites. Indeed, the Jacobites never intended that the focus of their cause should be in Scotland, they rightly recognised that England and London were the key to the British throne. For a while it seemed that Britain's historic enemy, France, would also play a significant role in the rebellion.

To distract Britain from intervention on the continent, France had been only too happy to support the Jacobites (taken from the Latin *Jacobus* or 'supporters of James', i.e. James Stuart). Between 1688 and 1745 there were fourteen major wars in Europe, many involving Britain. The largest French force to ever come within striking distance of Britain consisted of some 6,000 French troops supporting James Stuart in 1708. After reaching Scotland, it fled in the face of the Royal Navy. The lesson of the subsequent 1715 rising was that without external support, either foreign or national, Stuart restoration was impossible.

In 1722, the British Army adopted the famous flintlock Brown Bess or Tower musket. Arguably the best weapon in Europe, it stayed in service until the late 1830s. The seventeenth-century smoothbore matchlock musket had proved inaccurate and unreliable and, as a result, the more reliable flintlock slowly became more widespread in European armies.

The aftermath of Britain's involvement in the War of Spanish Succession, 1701–14, dragged it into the War of Austrian Succession of 1740–48. The former had seen Spain lose her empire, particularly her Italian possessions, and the latter resulted in the dismemberment of the Habsburg Empire. France, in particular, wanted the Austrian, formerly Spanish, Netherlands. Nonetheless, French aspirations received a setback when they were defeated at Dettingen in Belgium in June 1743, by an army composed of English, Hanoverians and Hessians, under King George II. Interestingly, the French failure was partly attributed to them fixing their bayonets too soon, thereby cutting their rate of fire. It is notable that, after the Peace Treaty of Utrecht in 1713, for almost the next thirty years infantry tactics were similar to those used by Britain and the Dutch United Provinces during the War of Spanish Succession.

Initially the French had earmarked 15,000 men to support Prince Charles Edward Stuart's rebellion in 1744, with 12,000 men landing in Kent and 3,000 in Scotland. The plan had been for a direct attack on London. Against this, the British Army could only muster 11,500 men in England and 2,800 men in Scotland. The situation was such that the British Government felt obliged to request the assistance of 6,000 Dutch troops. Fortunately, bad weather wrecked the Jacobite fleet and the War of Austrian Succession was of far greater concern to France. Austria, Britain, the Dutch Republic and Saxony concluded an alliance in January 1745 to counter Prussia and France. The Dutch were understandably concerned about French intentions in the Austrian Netherlands.

Britain then suffered a defeat at French hands at Fontenoy on 11 May 1745. This led to France's successful occupation of the Austrian Netherlands with the surrender of Ghent, Bruges and Ostend. King George II was in Hanover and the British Army in Flanders was in disarray – the time was ripe for another Jacobite rising. Yet France had lost interest in the enterprise and although keen to see Britain diverted, the French were now not overly anxious to supply the Jacobite cause with troops. Instead, they were keen to pursue their conquest of the Austrian Netherlands, followed by war with the Dutch Republic. In fact, the 1745 rebellion was to serve French interests for it caused a British withdrawal from the Low Countries.

Prince Charles Edward Stuart, known as the 'Young Pretender' (or even better, as 'Bonnie Prince Charlie'), resolved to act regardless. He decided that he might manage to force France's hand by commencing the rising without them.

Lacking French support, many Jacobites felt the venture was folly, so 1745 was to be make or break for the cause.

The prince was 24 years old when he arrived in Scotland on 23 July 1745 to claim the British crown. He landed at Eriskay with only seven supporters. Indeed, Charles Edward nearly came to grief before he had even landed. His little fleet, the *Doutelle* and the *Elisabeth*, the latter a sixty-four-gun warship carrying 700 men of the Irish Brigade, bumped into HMS *Lion* on 5 July. A four-hour naval engagement took place between the *Lion*, a sixty-gunner, and *Elisabeth*, about 100 miles west of the Lizard. HMS *Lion* came off worst and was forced to retire to Plymouth. Even so, the *Elisabeth* was so badly damaged that she was forced back to Brest, not only with Charles' professional core of 700 men, but also 1,500 muskets and 1,800 broadswords.

The government's control of the sea was to play a crucial role in denying the rebels much-needed regular French and Irish troops from the continent. Also, London was now alerted to the renewed Stuart threat. Shortly after this, King George II's third son, William Augustus, Duke of Cumberland, commanding the British forces in the Austrian Netherlands, was ordered to prepare to send troops to England immediately. He withdrew his forces safeguarding Antwerp in order to keep his lines of communication open. Intelligence reaching London indicated that Charles Edward had landed with up to 3,000 men and 2,000 weapons. Also 6,000 Spaniards and a squadron of warships were gathered off north-western Spain at Ferrol, possibly with a view to supporting an invasion.

Most of the Highland clans joined the young prince and he entered Perth. In the south, the English Jacobites, convinced that the French would follow up their successes in the Netherlands with an invasion of England, requested that 10,000 troops and 3,000 arms be sent to Essex in August –

nothing happened. In Scotland, General John Cope, commanding the government forces at Stirling, withdrew to Dunbar. On 21 September 1745 Charles, with 2,400 men, marched on Preston to attack Cope's forces. The following day, Cope was severely defeated at Prestonpans with the loss of 300 killed and 1,600 prisoners. Although the British regiments shamefully fled before the blood-curdling Highlanders' charge, most of them were inexperienced soldiers with a poor commander. Thus ended the effective Hanoverian opposition in Scotland, except for a number of troublesome fortifications.

Although the Austrian War of Succession continued to drag on, the battle at Prestonpans forced George II to summon more troops from the continent and withdraw from the Low Countries. Britain's ally, Charles Emmanuel III of Sardinia, was defeated on 27 September 1745 at Bassignano by a Franco-Spanish force. Austro-Sardinian troops were to gain their revenge less than a year later, on 16 June 1746, when they defeated the Franco-Spanish Army at Piacenza.

In the meantime, the Jacobite Army began its march south towards London, but the clansmen got cold feet and Charles' alarming offensive came to a halt. By 5 December the Jacobites had reached Derby, but support was not forthcoming. The Duke of Cumberland was barring the road at Lichfield with 7,000 redcoats, while General George Wade with 6,000 Dutch at Newcastle could cut the rebels off at Carlisle. Charles' forces only mustered 5,000 foot and 500 horse. The mathematics of the situation was apparent, to some, the Jacobites should have acted in 1744, but now it seemed too late.

Waiting in Scotland were welcome reinforcements. The Duke of Perth had landed with his own regiment, the French Royal Scots, and finally elements of the Irish Brigade, French mercenaries or 'Wild Geese'. In total, he managed to muster for Prince Charles some 4,000 soldiers. With no rising in England or any signs of a French invasion and with the British Government gathering 30,000 troops, the prince decided to withdraw and collect these much-needed reserves. On 6 December the majority of clan opinion forced Charles to move back north, thus away from the English throne and rendering his campaign purely defensive in Scotland.

By 17 December the rebel artillery and 500 Macdonalds were a day's march behind the Jacobite Army. Some government militia got between the main body of the rebels at Penrith and the rearguard under Lord George Murray at Clifton. The brief skirmish resulted in Murray calling for help. The following day, on 18 December, Murray successfully ambushed Cumberland's 4,000 redcoats on Clifton Moor.

The Jacobites moved unmolested on to Carlisle, where a garrison of 300–400 was left. On the 20th the prince divided his army and marched into Scotland. Ten days later, Carlisle surrendered to Cumberland. In the south, the government feared a large-scale French invasion, so having seen the Jacobites out of England, Cumberland was recalled to command the anti-invasion forces. Lieutenant General Henry Hawley was left to pursue the rebels.

Prince Charles occupied Glasgow on 3 January 1746 and then marched on Stirling, the army once again dividing. One column went via Kilsyth to Bannockburn with Charles, the other with Lord Murray via Cumbernauld to Falkirk. When the reinforcements reached the prince, his command swelled to 9,000 – the most men he was ever to attract. On 8 January the rebels took Stirling and besieged the garrison in the castle.

Lieutenant General Hawley in Edinburgh was gathering all available government forces, which grew to fourteen battalions of foot and three regiments of dragoons. His second-in-command was Major General John Huske, affectionately known to his men as 'Daddy'. Their ten pieces of artillery were under

Captain Archibald Cunningham, who Hawley described as 'a Sott' (drunkard). The troops he had scraped together could hardly be called first class. Hawley had been a major in the dragoons at the Battle of Sheriffmuir during the 1715 Jacobite Rebellion, he had also fought at Dettingen and Fontenoy. The man was determined to strike a decisive blow against the Jacobites and was convinced they would not stand against well-drilled cavalry. Hawley saw this as a prime opportunity to win favour at court. On 13 January 1746, Huske was despatched to search for the rebels, and two days later the main force set out to relieve Stirling Castle.

Lord Murray encountered government forces at Linlithgow on the 13th and the Jacobites began to concentrate at Bannockburn, except for 1,200 men left besieging Stirling Castle. Two days later, with their army embattled, many Scots no doubt hoped to repeat the victory of Robert Bruce in 1314. But by the 17th, when the Government Army had still not materialised it was decided to take the offensive.

Murray suggested occupying Falkirk Hill, 1 mile north of the government camp. Between 12 p.m. and 1 p.m. on 17 January 1746 the Jacobites headed for Falkirk. However, Lord Perth (James Drummond) had set off earlier with a body of men to march to the north of Torwood; this would be clearly visible from the enemy camp. The rest of the army marched to the south, by detouring it was hoped they would reach the ford over the Carron, 2 miles from Falkirk.

'I do and allwayes shall despise these Rascalls,' said Hawley, describing the Jacobite rebels, and he had good cause to despise them after the Battle of Falkirk. The Jacobites, lacking the training of a contemporary army and, in many cases, the weapons, had consistently defeated the government red-coated regulars. Falkirk was no exception, and was characterised by bad leadership on both sides.

At 5 a.m. that morning, Hawley reconnoitred towards Torwood, but he placed no picquets especially on the ford. At 11 a.m. Drummond's troops were spotted to the north and the government force stood to for 15 minutes. Two hours later, the main Jacobite body was seen south of Torwood. Word was sent to Callendar House where Lady Anne, wife of Jacobite Lord Kilmarnock, was entertaining Hawley to lunch. Hawley did nothing, distracted by the sumptuous meal and oblivious to what was going on around him. By 1.30 p.m. the rebels had crossed the ford at Dunipace Steps – they could not believe their luck that the crossing had not been contested, although the prince nearly lost his nerve. At 2 p.m., a second general alarm was sounded with the Jacobites only 1½ miles away. Huske, without any standing orders, again sent word to his wayward CO.

When Hawley finally arrived, he had 'the appearance of one who has abruptly left a hospitable table'. He immediately ordered the three dragoon regiments to seize the hill between the two armies. The slopes, which are intersected by folds and a deep ravine, rose from Falkirk to form a moorland plateau. The Government Army then began to advance as a storm blew up, beating the rain into their faces and dampening their powder as well as their enthusiasm. It is recorded that only one in four muskets would fire as a result of the weather. Most of the foot failed to reach the top before the Jacobites, nor was there any hope of hauling the government artillery up. En route six cannon were left stuck in the mud along 'Maggie Wood's Loan'. Fortunately, the rebels were forced to leave their artillery behind as well.

The Jacobites secured the element of surprise, descent and numbers and they forced Hawley to fight on broken ground completely unsuitable for regular troops. They deployed with their front line consisting entirely of

Highlanders – three Macdonald regiments of Keppoch, Clanranald and Glengarry on the right, next a small battalion of Farquharsons. In the centre were the Mackenzies, Macintoshes and Macphersons, and on the far left the Frasers, Camerons and Stewarts of Appin. The second line consisted of Lord George Murray's Atholl Brigade, Lord Ogilvy's and Lord Lewis Gordon's Regiments, some Maclachlans and Lord Drummond's Regiment. The few horse, under Lord Pitsligo and Earl Kilmarnock on the left, on the right Lord Elcho, Lord Balmerino and Murray of Broughton's hussars formed a third line – in total some 360 men.

Murray took charge of the right and Drummond supposedly the left, but he was to be in no position to control it. Like Hawley, Charles had been neglectfully vague in issuing orders. The prince positioned himself in the centre of the horse with the French/Irish Picquets. The Jacobites were in

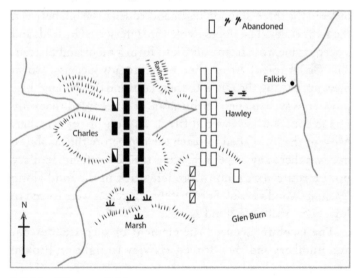

Falkirk, 17 January 1746.

a strong position; their left was protected by the ravine and their right by marshy ground.

The government redcoats deployed in two lines, the front consisting of (from right to left) Ligonier's, Price's, Royals', Pulteney's, Cholmondeley's and Wolfe's Regiments. The second line consisted of Barrel's, Battereau's, Fleming's, Munro's, Blakeney's and Howard's, with the Argyll militia placed on the far left and the Glasgow Volunteers on the right. Cobham's, Ligonier's and Hamilton's Dragoons were deployed on the left in front of the foot. The twelve battalions of foot each numbered about 400 men, the Glasgow Regiment 700, and the dragoons 200. Altogether, the Government Army mustered at most 8,500.

While the armies deployed for battle, a wit amongst the redcoats exclaimed, 'It's the Duke of Perth's mother!' as a hare darted between the two. A joke doing the rounds of the Government forces claimed that the lady in question was a witch who could turn herself into animals.

The dragoons remained inactive for fifteen minutes, and then were ridiculously ordered to attack the rebel line. Half the infantry were still forming up and a very unhappy Colonel Ligonier advanced without any support. Lord George Murray, with the Macdonalds, moved to meet the dragoons. By 3.50 p.m., when they were only ten to twelve paces apart, the Macdonalds opened fire and a ragged volley rolled as far as Lord Lovat's regiment of Frasers. Some eighty dragoons came crashing to the ground. Both Hamilton's and Ligonier's units wheeled about and fled, while Cobham's turned to the right and also rode off under fire.

Lieutenant Colonel Whitney and a group of men reached the Jacobites. He was killed, but some of Ligonier's troopers broke into the ranks of Clanranald's regiment. The Highlanders, however, were quick to counter this.

Throwing themselves to the ground they stabbed the dragoons' unfortunate mounts from beneath. Meanwhile, those fleeing carried away a company of militia from the Glasgow Regiment that was marching up the slope. The conductors of all but three of the gun teams, also struggling up the hill, fled. Hawley himself was carried away and although Huske remained in control it was an ill start to the battle. Cobham's demoralised men did rally, but fell back when the Irish Picquets advanced.

At 3.55 p.m. the government infantry and some horse advanced against the Camerons, Stewarts, Frasers and Macphersons on the Jacobite left. Most of the Highlanders' muskets were unloaded, having fired at Cobham's dragoons, and because the weather was so bad many had not bothered to reload. A few potshots were fired in answer to the government troops, and then the Highlanders drew their swords and fell on them.

The government infantry were quickly driven back down the slope. The rest of the Highlanders broke ranks to pursue the dragoons and fell on the government first line. Lord George found he had lost control as much of the army charged. The Jacobites were met at first by an uneven volley, after which the government first line simply gave way in the face of the blood-curdling charge. The second line did not even bother to fire, apart from Barrel's regiment the line disintegrated, while of the first, only parts of Ligonier's and Price's remained, protected by the ravine. These units bravely fired into the flanks of the attacking rebels.

Drummond refused to believe the regular redcoats had collapsed so easily, 'Surely this is a feint!' he cried. Believing it was a trap, he called off the pursuit. Both flanks of the Jacobite Army were screened from the other by the uneven ground, now the lack of communication and proper command was to hamper them severely, resulting in confusion and a lost opportunity.

Murray, on his wing, tried to rally his scattered men in order to cut off Hawley's retreat. The Highlanders, how-ever, felt the battle was over, the redcoats had fled – also, crucially, no bagpipes were to hand to rally them with. On Drummond's wing, some of his men thought the battle had been lost and many wandered from the field. In the centre was a gap which Charles and the Irish Picquets moved to fill, but the second line had added to the chaos by joining the first's charge, leaving only the Atholl men. The Jacobites had lost the chance of inflicting an even greater defeat on Hawley, and crucially, the bulk of the government troops were to escape.

It was now growing dark as the battle ended and, ironi-cally, most of Hawley's men were fleeing eastwards, while some of the Jacobites were withdrawing westward. The fact remains that Hawley had been disgracefully beaten; although at the time some argued it was indecisive because some of his units were still on the field.

The Battle of Falkirk lasted some twenty minutes, in which time the Jacobites lost about fifty killed and about seventy wounded. A government official return showed twelve offic-ers and fifty-five ranks killed, with 280 wounded and missing, and 600 prisoners of war. Other figures put the losses at six-teen officers and 300–400 men. Falkirk was occupied, and the government forces' tents and all their ammunition and some wagons were captured.

A contrite Hawley wrote to Cumberland, 'Sir, My heart is broke … suche scandalous cowardice I never saw before.' Five officers were court-martialled and cashiered, including Captain Cunningham, who attempted suicide, while thirty-one dragoons were hanged for desertion and thirty-two foot were shot for cowardice. Hawley himself escaped censure, although it is a little unfair to blame his defeat purely on the fibre of his men. It was a lucky victory in a remarkable string of them for the Jacobites, considering that most British rebellions were always doomed.

At this point, Charles should have acted decisively against local government forces. Instead, Murray was irritated by the prince who, suffering a heavy cold, dallied with his lover Clementine Walkinshaw. 'I am just ready to get back on horseback in order to make you a visit, but have been over-persuaded to let it alone by people who are continually teasing me with my cold,' wrote the prince to Murray. It was not until 1 February that he lifted the siege of Stirling Castle and moved on Inverness.

Murray and the others prevailed upon the prince to withdraw to the sanctuary of the Highlands. Such a move could only damage morale further and isolate the Jacobites from external help. Charles was faced with a fait accompli. His so-called friends on the continent did not care about his plight.

In the meantime, the Duke of Cumberland had gathered together all the regular troops he could muster and was marching in pursuit. While at Aberdeen, Cumberland's redcoats practised a new bayonet drill designed to counteract the Highlanders' charge. The problem was that the rebels' targes (shields) were deflecting the redcoats' bayonets too easily. The drill was that each man thrust to his right, thereby catching the Highlander as he raised his claymore sword. After six weeks Cumberland was confident that his men would not be caught out again.

The Jacobites continued foolishly to retire northwards and by 15 April were at Culloden, 4 miles east of Inverness. Cumberland, though, was only 12 miles away to the northeast at Balblair. The Jacobite Army, although tired and hungry after all its manoeuvring, deployed on Culloden Moor to face the expected attack, but Cumberland did not move, for it was his 25th birthday and he allowed his troops to rest.

Prince Charles, against his staffs' wishes, pressed for a daring surprise night attack on the duke's camp. The whole operation was a dismal failure, the army became divided and the prince, bringing up the rear, had no control over events. Lord Murray, who was leading the first body, sensibly decided to call the attack off and Charles was furious at the shambles. As the dejected army plodded back to Culloden, 2,000–3,000 men dispersed in a desperate bid to find food.

Early the following morning, Cumberland received news of the rebels' abortive attack and decided not to give them any time to rest. At 5.15 a.m. on 16 April 1746 the Government Army moved off from Balblair to give battle.

When word reached the prince that Cumberland was on the move, he made it clear that he was determined to stay and fight and seemed blind to the possibility of failure. The French emissary and leading Scots officers pleaded with him not to, but to retire to Inverness and await the rest of the army. Charles was adamant. Possibly he was still angry over having his orders slighted the previous evening, but now he had taken his fate into his own hands.

On paper the Jacobites had a force of 8,000 men, in reality, it was only between 5,000–7,000. When the alarm was given there were only about 1,000 men in the immediate area. Out of the Jacobites' pitiful force of 150 cavalry, the best were despatched to round up their scattered forces. Charles, mounting his horse, placed himself at the head of Lochiel's regiment and led his bedraggled army onto the moor (Drunmossie) south of Culloden House.

Cumberland's army, meanwhile, about 9,000 strong, was marching towards Culloden in three divisions, each of five battalions, with the 800 horse on the left, while the artillery and baggage followed behind. Amongst Cumberland's fifteen battalions of infantry were three regular Scots regiments, including the Campbells who had a lifelong hatred for the rebel Camerons.

Due to the route taken, two of the fourteen Jacobite cannon had to be left behind. Also, their army was drawn up further west than the previous day, and as a result the ground had not been reconnoitred. Therefore, because of insufficient numbers, the Jacobites were forced back into a very boggy area. They drew up about half a mile south of Culloden House, between the walls of Culloden Park and Culwinniac Farm. Only two lines were formed, the first of about eleven regiments and the second of six, including some 780 French/ Irish troops. In fact, less than half of Charles' army were true Highlanders, they made up the front line while the Lowlanders with the foreign troops formed the second.

The twelve cannon, manned by inexperienced gunners, were dispersed in batteries along the front line, with a four-gun battery in the centre. The few horse formed a reserve behind either wing. The Duke of Perth commanded the left, Lord John Drummond the centre, and Lord George Murray the right. Unfortunately, a squabble broke out between the Macdonalds and the Atholl brigade – the Macdonalds claimed it was their turn to fight on the right, although in the end the disgruntled Macdonalds were forced into line on the left.

Cumberland wrote proudly:

Sure never were soldiers in such a temper. Silence and obedience the whole time and all our manoeuvres were performed without the least confusion … It was pretty enough to see our little army form from the long march into three lines twice on our march, and each time in 10 minutes.

Cumberland's army formed up in three lines, the first two consisting of six battalions each, with three in the third acting as reserve. The horse and dragoons were placed on the wings. This done, the army moved forward to the sound of the drums and the Campbells' pipes. By 11 a.m. the two armies were 2½ miles apart and the government forces halted when the gap was 2 miles. Cumberland could now see that there was a space between Culloden Park and the Jacobite left. Also, the rebels' left was further from his line than their right. The Jacobites, though, fearing encirclement on their right flank, moved two battalions to guard the farm walls.

They faced each other for the next hour, but just after 12 p.m. the north wind began to blow rain into the faces of the clansmen. The government troops were unaffected, as it was blown against their backs. The Highland charge was met by a fusillade from four of Cumberland's battalions, as well as his artillery. Due to the intensity of this fire, those Mackintoshes who were not shot down veered to the right. As a result, the right-wing regiments (the Camerons, Stewarts and Atholl) were forced to bunch up into a tightly packed mass. While the Highlanders attempted to charge, Cumberland's gunners began to fire grapeshot, and the Campbells enfiladed their right flank and Wolfe's regiment lining the wall.

The Jacobites were so closely packed that only Barrel and Munro's regiments received the full impact. The Highlanders had been ordered not to throw away their firearms, but to use them to soften Cumberland's line. Many clansmen, however, were so bunched that few could actually fire, while others who were desperate to get to grips had thrown down their muskets.

Munro's regiment held their fire until the Highlanders were within 30 yards. A loud volley brought the whole front rank of the clansmen crashing down. Although the government

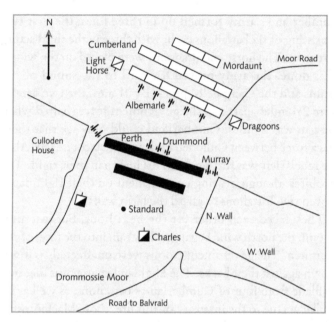

Culloden Moor, 16 April 1746, 11 a.m.–12 p.m.

troops had been instructed to thrust at their right-hand opponent, thus against the sword arm and not his shield, the impact carried Barrel's regiment backwards as the Jacobites burst through their ranks. The gun crew between Barrel and Munro's regiments held their fire until the Highlanders were only 2 yards away. At point-blank range, two barrels of grapeshot mowed down great swathes of the Highlanders, reducing their bodies to a bloody mess.

The Grenadier company on Munro's left was simply swept away. From Cumberland's second line, two regiments, Bligh's and Sempill's, were moved 50 yards forward to check the breakthrough. Of the 500 Jacobites who had managed to reach this second line, most were cut down. In sheer desperation, some of the clansmen frantically hurled stones as their casualties mounted. Resistance lasted several minutes as the attack collapsed and turned to rout.

Meanwhile, on the left the Highlanders had still not moved – two regiments did not even possess swords. Finally, though, the Macdonalds advanced 100 yards. Three times they made as if to charge, brandishing their swords and yelling. Lord Drummond even tried to draw the government regiments' fire in order to get them to waste a volley, but failed.

Around 1 p.m., the Jacobites' four-gun battery opened fire, aiming it seems at Cumberland. One shot just missed him and another whizzed by to kill two men behind. The government's ten 3-pounders, placed in pairs along their front line under Colonel Belford, were directed to return fire. Ranging in on a party of horsemen behind the second Jacobite line, the prince soon found himself under artillery fire. His groom was killed and Charles was urged to move, which he did, to a poor observation point.

The government guns, manned by trained gunners, were soon firing quickly and accurately at the Jacobites' infantry and cannon, felling men left, right and centre. Their central battery was quickly silenced. The clan regiments were badly bunched, up to six ranks deep; men were still arriving and the shot was simply ploughing through them. Huge gaps were appearing as the Highlanders' dead and wounded began to pile up.

Cumberland noted:

They began firing their cannon, which was extremely ill-served and ill-pointed. Ours immediately answered them which began their confusion. They then came running on in their wild manner, and upon the right where I had placed myself imagining the greatest push would be there, they came down there several times within a hundred

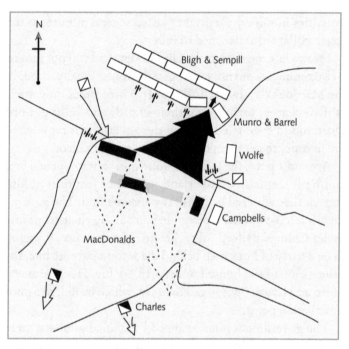

Culloden Moor, 16 April 1746, 12.30 p.m. to 1.45 p.m.

yards of our men, firing their pistols and brandishing their
swords, but the Royals and Pulteney's hardly took their
fire-locks from their shoulders …

Cumberland was in no hurry to move. He had suffered few
casualties and his guns were doing maximum execution. The
government bombardment lasted for thirty minutes before
anything happened. The Highlanders were being galled
beyond endurance and their fighting spirit was draining fast.
Lord Murray requested that the prince order the attack, but
it was too late. The Duke of Perth with the Macdonalds was
800 yards from Cumberland's line, whereas the right was

only 500 yards away. Knee-deep in water, suffering from the withering artillery fire and with such a long distance to cover, the Macdonalds would not move. However the centre, led by the Mackintoshes, with a loud cheer, swept forward followed by the right.

The Macdonalds, upon viewing the confusion on their right and the collapse of the attack, suddenly fled. Vainly Drummond tried to halt them, but Cumberland's right began to advance and fired a volley to see the rebels off. Some gentlemen and officers of the Macdonalds shamed by their brethren, charged Price's regiment, only to be cut down. Lord Murray, having fought his way back to the second line, ordered forward Perth and Glenbucket's regiments to cover the fleeing Macdonalds. Prince Charles was completely dismayed by this disaster, refused to believe all was lost and had to be forced from the field. He knew now that he had lost all chance of ever gaining the English crown.

The Campbells on the right volleyed the broken Jacobites as they fled. A Gordon battalion faced them, but the dragoons outflanked the rebels and forced back Fitz James' horse. On the Jacobite left wing, Kingston's light horse pursued the Macdonalds, who took with them the second line. Some of the professional French units and Irish Picquets held their ground, driving off the light horse, who then moved to the centre to link up with the dragoons. Cumberland's foot now advanced, while Lord Murray rallied several regiments and moved off southwards. The shattered Highlanders were pursued as far as Inverness.

The actual fighting lasted just a little over 40 minutes and, of the 14,000 men present, it is believed that only about 3,000 were actively engaged. At least 1,200 Jacobites were killed, many of them during the rout. The government troops suffered 310 casualties – fifty killed, 259 wounded and one missing.

Prince Charles fled to the Highlands and eventually to France, but not before he became a legend, evading Major General John Campbell of Mamore and 2,000 redcoats. As for Cumberland, he allowed his troops to sully the name of the British Army in the reprisals that followed Culloden Moor.

The Battle of Falkirk was the final chance for the Jacobites to aggressively press home their claim in England. Instead, they did not act decisively and Falkirk turned into a wasted opportunity, laying the grounds for defeat at Culloden. Charles Edward failed to reoccupy Edinburgh; that, and their failure to take Stirling when the garrison was on the point of capitulating, saw Jacobite morale fall and desertion became a serious problem. While the French abandonment of the Jacobite cause contributed to its downfall, England would not have tolerated a change of dynasty brought about by force of French arms. Any French support for the Stuarts had always been designed to serve French interests. France's failure to mount an invasion in the winter of 1745–46 was easily compensated for by their capture of Brussels on 20 February 1746 – which had been partly facilitated by British forces being recalled to England.

Falkirk was greatly affected by two factors. First, the terrain, which divided the Jacobite command and put the government forces at a disadvantage. Second, morale was a key factor, the government troops were bedraggled, with the rain in their faces and wet muskets, and the highly strung Jacobites poised for that single charge which the redcoats so greatly feared. In contrast, at Culloden the Highlanders' morale was poor, even at the start of the battle. Culloden was decided simply by the government forces' firepower. Cumberland's army was not subject to any preliminary bombardment and was able to give well-disciplined fire when necessary.

In 1066 England faced invasion by the Normans and the Vikings, resulting in a two-front war that the Anglo-Saxons could not win. (Author's collection)

The hard-won Saxon victories at Fulford Gate and Stamford Bridge ultimately contributed to King Harold's defeat at Hastings. (*Battle of Stamford Bridge* by Peter Nicolai Arbo, 1870)

While the Saxons and Vikings rode to war and dismounted to fight, the Normans fought from horseback. (Bayeux Tapestry, 1070–80)

The Battle of Homs in 1281 sounded the death knell of the Crusader kingdoms in the Holy Land. (*Histoire des Tartares, Hayton of Coricos*, fourteenth century, Bibliothèque Nationale de France)

Crécy in 1346 stands out as a triumph of British arms during the Hundred Years War. The longbow proved a great equaliser when confronting mounted armoured French knights. (*Froissart's Chronicles*, fifteenth century)

A rather dramatic Victorian depiction of Bosworth in 1485. The use of gunpowder on the battlefield marked the start of the pike-and-shot period. (*Battle of Bosworth Field* by Philip James de Loutherbourg, 1857)

Lord Stanley presents the crown to Henry Tudor at Bosworth. Richard III was the first English king to be killed in battle since 1066. (*Cassell's Illustrated History of England*, 1858)

Hernán Cortés, secretary to the governor of Cuba, brought the Aztec Empire down through a mixture of audacity and guile. (*Prescott's History of the Conquest of Mexico*, 1843)

The Conquistadors did not only use gunpowder and horses to defeat the Aztecs; they also employed coalition and biological warfare. ('Tlaxcalan Cortez' facsimile, 1800s)

The Aztec island fortress of Tenochtitlan held out for eighty days before the Spanish secured victory in 1521. (*Coxhead's Romance of History*, Mexico, 1909)

Jean de Valette, commander of the Knights of the Order of St John and defender of Malta in 1565. Naval and siege warfare characterised the fighting in the Mediterranean during the 1500s. (Laurent Cars, 1725)

The elite Janissaries were the sultan's shock troops who led the attack on Malta in 1565. (Chamberlain of Sultan Murad IV with Janissaries)

Turkish artillery pounding Malta's defences. The use of gunpowder in siege warfare led to a radical redesigning of fixed fortifications and rendered the traditional castle obsolete. ('Attack on St Michael' by Matteo Pérez d'Aleccio, 1582)

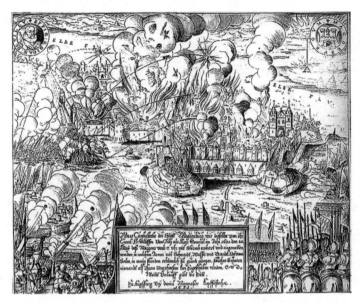

The German Protestant city of Magdeburg under siege in 1631 during the Thirty Years War, which tore Europe apart. Its refusal to surrender resulted in the inhabitants being put to the sword. ('Sack of Magdeburg' by Matthäus Merian, seventeenth century)

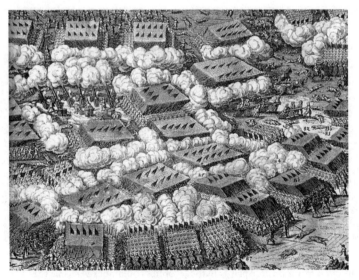

The sacking of Magdeburg was followed by a victory for the Protestants led by Swedish King Gustavus Adolphus at Breitenfeld. Cavalry, cannon, muskets and pikes ruled the battlefield. (Engraving, seventeenth century)

The Jacobites' remarkable victory at Falkirk on 17 January 1746 was not followed up, leading to defeat at Culloden where they once more pitted the broadsword against the musket and bayonet. (Contemporary engraving, eighteenth century)

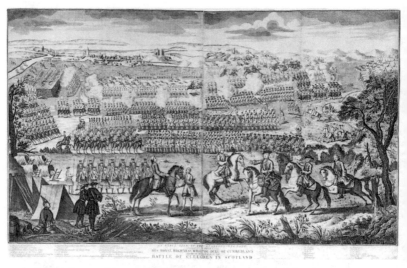

At Culloden Moor on 16 April 1746 the Jacobites could not overcome the Government forces' superior tactics and discipline. ('Battle of Culloden in Scotland', contemporary engraving, eighteenth century)

Napoléon.

At Waterloo in 1815 Napoleon brought about his own demise, for it was he who committed the fatal error of dividing his army at the deciding moment. (Author's collection)

Marshal Blücher was 'damnably mauled' at Ligny but recovered sufficiently to fight at both Waterloo and Wavre. (Maxwell's, *The Life of Wellington*, 1900)

Marshal Grouchy had none of Ney's dashing flair – ignoring his officers' pleas to march to the sound of the guns at Waterloo, his actions at Wavre would forever haunt him. (Maxwell's, *The Life of Wellington*, 1900)

Grouchy, with almost a third of the French Army, fought a futile battle at Wavre. For forty-eight hours Thielemann's Prussian corps tied up the French right wing, successfully preventing two much-needed French corps from reaching Napoleon. (Author)

In 1879 a Zulu army armed largely with just spears overwhelmed a modern British force armed with breech-loading rifles at Isandhlwana. The British foolishly split their army. (Ian Knight)

The ominous looking Isandhlwana Hill – geography played a key part in the battle, helping to conceal the Zulus. (Ian Knight)

The First World War heralded the onset of armoured and aerial warfare. British and German tanks fought the very first tank battle at Villers-Bretonneux in 1918. (Author)

A cumbersome German A7V tank – the very first Panzers were very crude affairs. (Author)

In 1940 the British Isles braced itself for invasion by Hitler's Nazi Germany. (*Portsmouth News*)

RAF Fighter Command fought the Battle of Britain and repulsed the Luftwaffe, but Hitler's seaborne invasion never came. (Author)

Hitler hoped his Blitz would pulverise Britain into submission, but it did not work. Nonetheless, the Allies adopted this same strategy against Nazi Germany for the rest of the Second World War. (Author)

On 6 June 1944 the Allies launched Operation Overlord, the largest amphibious operation in history, which came ashore in Normandy. (US National Archives & Records Administration (NARA))

D-Day required specialised transport vessels to deliver tanks and other equipment onto the invasion beaches. (NARA)

Although the US Army beat the British and Canadians over the mighty Rhine in early 1945, Field Marshal Montgomery insisted on conducting a major set-piece battle to secure the river. (NARA)

British Buffalos moving up during Operation Plunder, launched on 24 March 1945. The Rhine crossing was the last major amphibious operation of the war in north-west Europe. (War Office photograph)

Landing craft used on D-Day were also employed to get tanks across the Rhine. (NARA)

In 1968 American air support and resupply played a key role in the Battle of Khe Sanh. (US Department of Defense)

North Vietnamese anti-aircraft measures almost completely cut off the American firebase at Khe Sanh. (North Vietnamese Army)

Air drops proved instrumental in sustaining the garrison at Khe Sanh. (US Department of Defense)

In 1991 most of Saddam Hussein's armour thrown into the attack at Al-Khafji was destroyed by Coalition air strikes. His air force was notable by its absence. (US Department of Defense)

Saudi Arabian armed forces deploying in 1991. The Gulf War was a prime example of modern coalition warfare conducted on land, air and sea. (US Department of Defense)

Zhawar Kili Support Complex, Afghanistan

In 2001, Afghanistan was harbouring terror leader Osama bin Laden and Al-Qaeda, as well as hosting numerous terrorist bases such as this one. (US Department of Defense)

The fate of the terror bases in Tora Bora. (UK Ministry of Defence)

11

An Army Divided: Wavre 1815

In less than eight hours on 18 June 1815, some 200,000 men fought in an area of about 10 square miles to decide whether Emperor Napoleon Bonaparte would reassert his dominance of France and Europe. His ambitions were to be thwarted by his own poor performance and by a deficiency of flare and audacity from his marshals. In particular, Marshal Grouchy's complete lack of initiative and prosecution of the fruitless Battle of Wavre ensured Napoleon's fate at Waterloo. Grouchy allowed the Prussians to get between himself and Napoleon; the net result was that the pressure was taken off Wellington's sorely beleaguered army at a crucial moment in the battle.

Napoleon's escape from Elba in March 1815 at the head of just 1,500 men was an astounding comeback after his defeat the year before. He had been forced to abdicate after twenty-five years of war and the Bourbon monarchy restored under Louis XVIII. Europe's response to Napoleon's triumphal return to Paris was to declare him an outlaw and mobilise the Seventh Coalition against France since 1792. He did not want war, he even wasted time trying to appease the French faction that had removed him in 1814, but the allies would not treat with Napoleon. Time was against him, France was worn out by the Revolutionary and Napoleonic Wars. Even his loyal marshals were reluctant for renewed hostilities. His hope was that if he could swiftly defeat the British in Belgium he could trigger the collapse of the Seventh Coalition.

Napoleon planned to have 800,000 men under arms by October 1815. In the meantime, he gathered 124,000 troops of the newly constituted *Armée du Nord* (Army of the North) to attack Belgium. This force was largely veterans and is considered one of the best armies he ever commanded. In particular, it had a high proportion of cavalry and artillery. In sharp contrast, his selection of senior military leaders was poor and was to have a fundamental impact on the coming campaign. Instead of using such proven men as Marshals Suchet, Davout, Murat and Berthier, his able chief of staff (who was now dead), he ended up with Soult as chief of staff, and Ney and Grouchy commanding his groups of corps. Marshal Mortier was supposed to command the Imperial Guard, but he was ill and replaced by a lieutenant general. Napoleon had five major infantry corps under the experienced Generals Drouet D'Erlon, Reille, Vandamme, Lobau and Gérard, plus four smaller cavalry corps and the guard.

Two allied armies gathered in Belgium to counter Napoleon's unwelcome return to power – an Anglo-Dutch force under the Duke of Wellington and a Prussian Army under Field Marshal Prince Gebhard Blücher von Wahlstatt. Significantly, these forces were stretched over almost 100 miles, with the Prussians to the east and the Anglo-Dutch to the west, the strategic Brussels road lying between them. Wellington had some 94,000 men of varying quality and numerous nationalities, all speaking different languages. Notably, 9,000 Belgian troops had once been French citizens. The 117,000 Prussians consisted largely of ill-equipped *Landwehr* battalions and included 14,000 Saxons, also of dubious loyalty.

The smoothbore flintlock muskets of the day were broadly comparable. British muskets, such as the Brown Bess, fired a larger ball and were generally manufactured to a superior standard than French weapons. Although French muskets were lighter and easier to handle, the British ones had better stopping power. Prussian muskets had a crude finish like the French, but were of a similar size to the British.

Rifles were used in limited numbers by British, Dutch, German and Prussian units. To enable the slow loading musket and riflemen to protect themselves against cavalry, their weapons were equipped with detachable socket bayonets. These were a vast improvement on the early plug bayonet which blocked the barrel.

Artillery, by the Napoleonic Wars, was much more mobile and consisted of cannon and howitzers that fired round shot and shells respectively. Below 400 yards, cannon could fire devastating canister or case shot, this consisted of a metal cylinder full of musket balls that sprayed out of the muzzle and was particularly effective against formed bodies of troops. Howitzers fired a spherical shell that exploded in the enemy's midst.

Due to the Seventh Coalition, Napoleon was forced to commit valuable resources to five other widespread theatres of operation. There were 210,000 Austrians gathering on the Upper Rhine, 150,000 Russians on the Central Rhine and 75,000 Austrians facing the Riviera. To counter them, he had just 23,000 men covering the Upper Rhine, 15,500 covering the Italian frontier, 8,000 facing Spain and 10,000 dealing with the Royalist insurrection in the Vendée.

War broke out first in Italy. Napoleon's only external ally was former Marshal Murat, now the King of Naples, who could field 40,000 men. Before his escape from Elba, Napoleon instructed Murat to prepare the Neapolitan Army, but warned him to avoid open hostilities. Napoleon would not need to waste troops securing France's south-eastern border while Murat was distracting the Austrians in northern Italy.

Despite his orders, Murat acted too quickly, overwhelming 6,600 Austrians on the River Panaro on 3 April. He was then obliged to divide his forces in order to contend with more approaching enemy troops. His depleted army of 28,000 was subsequently repulsed at Tolentino on 2 May 1815 by just 11,000 Austrians and he fled to Corsica hoping to offer his services to the emperor. Napoleon was so incensed that he would not even see him, let alone offer him a command. Again, Napoleon did himself a disservice, for Murat was the best cavalry general he ever had and his presence at Waterloo could have made all the difference.

Wellington's understandable concern was that Napoleon would thrust westward to cut him off from the Channel ports. However, Napoleon's intention was to go for broke by driving a wedge between the two allied armies and then defeat them piecemeal. He knew that the British forces were routed from Brussels and the Prussians from Namur. Their junction would be conducted on the Namur to Nivelles road, which goes to Sombreffe and crosses the Charleroi to Brussels road at Quatre Bras. Napoleon planned to place his army at Quatre Bras and Sombreffe by the second week of June. If both his opponents fell back without a fight, he could march directly on Brussels.

Marshal Grouchy, initially given command of the cavalry reserves, was placed in command of the French right wing. Although he had fought in numerous battles, his appointment was to prove unwise. Despite his military achievements he had never been part of Napoleon's inner circle and has been described as cautious and unimaginative. He had only received his marshal's baton on 15 April 1815 after suppressing a Royalist insurrection in the Midi. Having started in the artillery, then commanded infantry, he was principally a cavalryman. Grouchy had served in France, Germany, Italy, Spain and Russia. He had taken part in the 1806 campaign against Prussia, serving as general of dragoons, during which he acquitted himself well at Eylau and Friedland.

With the French 3rd and 4th Corps, Grouchy was ordered to clear any Prussians before Sombreffe. Outnumbered, General Zieten's Prussian 1st Corps withdrew. Marshal Ney, commanding the French left with the 1st and 2nd Corps, was to clear the Brussels road. Like Grouchy, he was slow in executing his role, he halted at Gosselies after driving the Prussians out, some 3½ miles from Charleroi, losing his chance of taking Quatre Bras. If this had happened, Wellington would have had to withdraw and, with the Prussians out the way, eventually the Anglo-Dutch forces would have been at the mercy of French superior numbers.

Wellington realised on 15 June that Napoleon was about to fall on Charleroi and belatedly ordered a concentration at Quatre Bras. Indeed, Napoleon's strategy was going to plan. On 16 June, just one day after crossing the frontier, he split the allies and struck the Prussians with such force at Ligny that it seemed they would not recover. Initially Napoleon and Grouchy thought they were advancing against 40,000 Prussians, but found themselves fighting 84,000. The Prussians were outclassed; giving battle on ground that Wellington had warned Blücher was not suitable. He had remarked, 'If they fight here they will be damnably mauled.'

Although their centre collapsed, the Prussian Army was not destroyed. Napoleon, forgetting about Lobau's reserve corps, demanded troops off Ney at Quatre Bras to finish the task. This resulted in D'Erlon's 20,000-strong corps marching backwards and forward like complete idiots, taking part in neither battle. At one stage Vandamme, on Napoleon's left, thought the column advancing on his rear must be the enemy, causing discord amongst his exhausted troops. The net result was that Napoleon was distracted and his plans dislocated. Even the Prussians thought the column was British and renewed their efforts.

This level of poor communication was to characterise the rest of the 1815 campaign. Regardless, Wellington was proved right and the Prussian Army suffered 16,000 casualties and lost 9,000 deserters. Blücher himself was unhorsed and injured trying to stem the French tide.

On the same day, at the same time, Napoleon was to lose a second opportunity at Quatre Bras. Marshal Ney fell on the small Anglo-Dutch force, but then withdrew. He had 16,000 men against an initial Anglo-Dutch force of just 7,000. However, 22,000 allied reinforcements arrived and by 5 p.m. Ney, having lost D'Erlon, found himself outnumbered and retired. The allies lost 3,463 killed and wounded in the fighting while the French acknowledged losses of 4,300. No one had impressed upon Ney that Ligny was the centre of gravity. If he had moved to help Napoleon and Grouchy at Ligny, the Prussians would have been completely routed. Instead, Ney wasted time fighting an unimaginative battle at Quatre Bras while his reserves were needlessly squandered marching between the two engagements.

On the face of it, Napoleon had every reason to feel pleased with the results. The allies had been kept apart, one had been seriously mauled and the other fought to a draw. Ligny was a major French success, although they had lost 11,500 casualties. Nevertheless, Ligny and Quatre Bras actually had the reverse effect of forcing the two allied armies together. Then, to compound matters, Grouchy failed to place himself between them.

After Ligny there was no prompt French follow-up and this break in contact was to have disastrous consequences for Napoleon. Some of Grouchy's cavalry were sent off: Lieutenant General Count Exelmans' 2nd Cavalry Corps north-east towards Gembloux, and Lieutenant General Count Pajol's 1st Cavalry Corps towards Namur. Prussian stragglers gave the impression that their army was indeed withdrawing on Namur. Fortunately for Wellington, Blücher marched north towards Wavre instead of south-east towards Namur, where his bases were. This made them closer together.

Marshal Grouchy, with 33,000 men and ninety-six guns of the 3rd and 4th Corps, almost a third of the entire French Army, was eventually ordered to pursue the retreating Prussians at 1 p.m. on 17 June. The roads were atrocious and Grouchy's progress was painfully slow, covering less than 10 miles, and he gave up at nightfall, stopping at Gembloux, 12 miles south of Wavre. He was to remain there until 11 a.m. on 18 June; he received orders at 4 p.m. on the 18th that simply confirmed his movement northward towards Wavre. Napoleon lost the chance to order Grouchy to reposition himself between Waterloo and Wavre.

Following Ligny, Wellington, with his flank exposed by the Prussian withdrawal, fell back to pre-scouted positions near Mont-St-Jean. Still fearful that the French would try to outflank him to the west and cut him off from his supply base at Ostend, he despatched 17,000 men towards Hal, 7 miles from Mont-St-Jean. Knowing that he was outnumbered, Wellington proposed to fight a purely defensive battle as he had done in Portugal and Spain during the Peninsula War. His intelligence was that Grouchy had been ordered to follow the Prussians with only 12,000–15,000 men, not twice that number. To conceal and protect his army, he straddled the low ridge that ran across his new position south of a village called Waterloo. In front, three strongpoints were formed by the Château Hougoumont on the right, the farmhouse of La Haye Saint in the centre and on the left by the villages of Papelotte, La Haie and Frischermont. Two flooded rivers also protected his eastern flank.

Unhindered by Grouchy, the Prussians took up positions around Wavre on 17 June, receiving word that Wellington had taken up his at Mont-St-Jean. The Prussian 1st and 2nd Corps, which had borne the brunt of things at Ligny, arrived non-combatant as their ammunition stocks were so low. They were not replenished until 5 p.m. and 3rd and 4th Corps did not arrive in the area until the evening. Wellington made it clear he intended to fight if he could rely on reinforcements from one Prussian corps. Blücher's chief of staff, General August von Gneisenau was reluctant to risk the Prussian Army so soon after its beating. Fortunately however, Blücher prevailed. He wrote to Wellington generously offering two corps and a promise that the first would start off at daybreak. It is quite possible that the single corps Wellington requested would have been insufficient and the day lost.

Gneisenau did not altogether trust Wellington, especially as at the Vienna Congress the duke had not supported Prussia's claim to Saxony. Furthermore, Gneisenau did not want his army strung out with Grouchy somewhere behind him and Napoleon in front, especially if Wellington did not hold his ground at Mont-St-Jean. Therefore, to buy time Gneisenau unhelpfully persuaded Blücher to send Bülow's 4th Corps, which was actually the furthest away.

Over breakfast on 18 June, Napoleon was warned that the Prussians were to rendezvous with Wellington before the forest of Soignies and would come via Wavre. In contrast, he assessed that after Ligny a junction of the allied armies would not be possible for two days. Grouchy's forces, continuing their leisurely pursuit, were supposed to have seen Vandamme depart Gembloux at 6 a.m. and Gérard at 8 a.m. However, Exelmans' vanguard was late and Vandamme wasted an hour and a half. By 11 a.m. Grouchy was breakfasting at Sart-a-Walhain.

At Waterloo, Wellington gathered about 68,000 men and 156 guns, facing 72,000 French, of whom almost 16,000 were cavalry, with 246 guns. Both sides had spent a miserable night being soaked by torrential rain. During the morning, Soult counselled Napoleon to recall at least part of Grouchy's force as he had done the day before. 'You think,' responded Napoleon contemptuously, 'because Wellington defeated you, that he must be a great general. I tell you that he is a bad general, that the English are poor troops, and that this will be the affair of a *dejeuner.*'

'I hope so,' replied Soult dejectedly.

Bülow's Prussian 4th Corps did not reach Wavre and begin crossing the River Dyle until 7 a.m. In theory, even allowing for the sodden roads, at the latest they should have covered the 8 miles from Wavre by about 2 p.m. However, an hour later a serious fire broke out in the town delaying them for two hours as they tried to put out the flames. Blücher and his staff left at 11 a.m., leaving orders for 2nd and 1st Corps to follow the 4th in that order. Wellington was going to have to hold on for much longer than he anticipated, indirect help would not arrive until 4 p.m. and direct help not until 7 p.m.

Napoleon was in a hurry and had no intention of trying to outflank the Anglo-Dutch Army. He sent a regiment of cavalry and battalion of infantry to his extreme right, behind Frischermont, deploying some of the men in the woods of Bois de Paris. His intention was to watch for the Prussians rather than Grouchy.

The Battle of Waterloo commenced with the French launching a series of unimaginative frontal attacks that were only distinguished by the bravery of those executing them. Papelotte, La Haie and Frischermont, held by Bernard of Saxe-Weimar's Nassau troops, were the first to feel French steel. Napoleon was ill and suffering a general malaise, so he abdicated much of the direction of the battle to Marshal Ney. The 3,500-strong garrison of Hougoumont tied up 14,000 Frenchmen, while La Haye Sainte was only taken after the valiant defenders ran out of ammunition.

When battle commenced at about midday, Grouchy was lunching on strawberries 15 miles away to the east at Sart-a-Walhain. He claimed the cannon fire was a rearguard action, but as the exchange grew louder it was clear something far more serious was taking place. A peasant informed him that the noise was coming from Mont-St-Jean, just 8–9 miles away. Some of his officers called on him to march on the sound of the guns, but he was resolved to continue his chase of the Prussians. General Gérard, commander of 4th Corps, even pleaded to go with just his force of 15,500 men; Grouchy countenanced none of it and would not divide his command.

Nonetheless, he had been aware even in the morning that Blücher's forces were shifting westward through the Bois de Paris towards Plancenoit. Despite this, Grouchy showed no initiative or inclination to impede the junction of the two allied armies. His orders centred on Wavre, and Wavre was where he was going. Also, it seems he expected at least part of the Prussian Army would turn and fight him if he became overextended. To compound his frame of mind, a messenger arrived from Exelmans reporting a strong Prussian rearguard covering Wavre.

No one knew it at the time, but a French tragedy was about to unfold. If he had acted then, Bülow's corps could never have marched on Napoleon's right and would have been drawn back eastwards. Grouchy was not supposed to fight a pitched battle against the Prussians, rather to assist Napoleon. If he had allowed Gérard to go he would have provided the *Armée du Nord* with reinforcements at a critical time, and if he had attacked Wavre with the rest of his force in a timely manner, Blücher would have been held up and Gérard could have intervened at Plancenoit at 7 p.m. Likewise, if Napoleon had summoned Grouchy late on the 17th, he could have reached Plancenoit by 2 p.m. on 18 June.

Indeed, a Prussian prisoner revealed to Napoleon that Blücher and 30,000 men were on their way to attack the French right. A message was sent to Grouchy at 1 p.m. including a note from Soult, 'General Bülow is about to attack our flank. ... therefore lose not an instant in drawing nearer to us and joining us, in order to crush Bülow whom you will take in the very act'. Grouchy, who thought he was chasing Bülow, did not get this order until after 5 p.m., by which time it was too late. If he had been instructed to move sooner, the outcome of Waterloo may have been swayed in Napoleon's favour. In the meantime, Napoleon was forced to move Lobau's 6th Corps, some 10,000 men and twenty-eight guns, to counter the approaching Prussian threat.

Exelmans, acting as Vandamme's advance guard, was attacked sometime before 2 p.m. at Mont-St-Guibert by a mixed Prussian force. Grouchy decided to outflank it, but the Prussians conducted a fighting withdrawal before Exelmans' 3,000 dragoons could cut their line of retreat. Vandamme was ordered to pursue them as far as Wavre then await orders.

At about 3 p.m. Napoleon received a despatch from Grouchy written at 11.30 a.m., indicating he was 8 miles from Wavre. Grouchy's report stated, 'The 1st, 2nd and 3rd Prussian Corps, under Blücher are marching towards Brussels ... This evening I shall have massed my troops at Wavre, and thus shall find myself between the Prussian

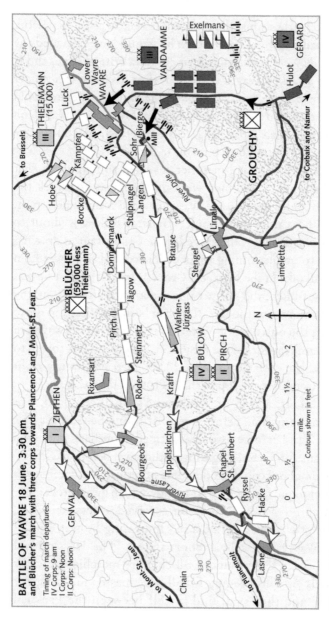

Wavre, 18 June 1815, 3.30 p.m.

Army and Wellington, who, I presume, is retreating before your Majesty.'

How wrong he was on all counts, and Napoleon knew it. By this time, Bülow's 4th Corps was almost on the French right flank. Grouchy's message simply confirmed Napoleon's greatest fear, that the missing Prussian 4th Corps was already at Chapelle-St-Lambert and that the others would be following. Knowing Blücher from two previous campaigns, Napoleon expected him to be marching post-haste with every available man towards Mont-St-Jean. The question was, where was Grouchy? Napoleon did not send a reply, for there was nothing to add to the despatch he had sent earlier.

Napoleon still had time to conduct an orderly withdrawal from Waterloo behind his cavalry, with Lobau's men as rearguard, but he was desperate for victory and fought on, continuing his attacks on Wellington. At 4 p.m. Bülow's corps arrived and elements of Lieutenant General Domain's 3rd Cavalry Division (detached from Vandamme's 3rd Corps) and Lieutenant General Subervie's 5th Cavalry Division (detached from Pajol's 1st Cavalry Corps), who were watching the woods near Frischermont found themselves under Prussian artillery fire. After being attacked by Lobau, the Prussians struck towards the village of Plancenoit, which was unoccupied. Lobau managed to get some of his reserves into the village just ahead of them. This was almost in the French rear, forcing their army to fold in half. Napoleon had been fighting Wellington on a 2½-mile front, his army now pivoted facing the Prussians, also on a 2½-mile front.

A messenger from General Thielemann at Wavre reached Blücher and Gneisenau, informing them that he was under attack and in danger of being overwhelmed. Gneisenau's response was, 'Let Thielemann defend himself as best he can, it matters little if he be crushed at Wavre, provided we gain the victory here.' There lay the heart of the matter – Wavre would be Napoleon's undoing.

Napoleon, appreciating that he was currently only engaging one Prussian corps, which Lobau was successfully holding off, fought on. He knew that if Wellington's centre gave under the massed French assaults, the duke would have to retreat. Only then would the Prussians break off their attacks, for they would become pointless. Furthermore, they would have to withdraw for fear of being trapped between Grouchy and Napoleon's main force. Although outnumbered three to one, Lobau's men initially stood firm, then withdrew towards Plancenoit. Between 4.30 p.m. and 5 p.m. Bülow attacked Plancenoit. Indeed, Wellington heard artillery fire coming from the south-east at about 4.30 p.m., informing him that the Prussians were supporting him indirectly somewhere. French resistance around Plancenoit held for two hours and when the Prussians finally took the village, the Young Guard threw them out driving them back for over a mile. Some 14,000 French troops were now tied down by the Prussians instead of committed to the final effort against Wellington.

Just 10 miles away, an equally desperate, but futile, struggle was fought at Wavre. By 4 p.m. Grouchy's troops were in action near Wavre against Thielemann's 3rd Corps. Blücher had intended Thielemann to advance on Mont-St-Jean, but only two brigades had crossed the river when French cavalry were sighted. Indeed, Vandamme's French 3rd Corps arrived just as the Prussians were leaving, catching Colonel von Zeppelin's two fusilier battalions. Thielemann knew he must thwart the French taking the bridges over the River Dyle. Under his instructions, they were barricaded and the houses loop-holed for muskets to fire from.

The River Dyle was unfordable, making the two stone brides at Wavre and the wooden bridges at Limale, Bierges and Basse-Wavre vital. The Prussians had about 15,000 men on a 3-mile front. Thielemann's orders were to hold Wavre, but if forced to withdraw he must lead Grouchy away from the main Prussian Army. Blücher had stated to Gneisenau, 'it matters not if Thielemann and his men stand and die to the last man, as long as they keep the French away from Napoleon.' Thielemann deployed his 10th Brigade behind Wavre, 12th Brigade behind Bierges and the 11th the other side of the Brussels road. The bridge at Bierges was barricaded and the mill fortified.

Thielemann and Grouchy's actions at Wavre were key to the unfolding events to the west. Had Thielemann abandoned the town, Grouchy could have had time to take stock of what was happening elsewhere. Even a token force, particularly cavalry, sent westward might have drawn some of the pressure off Lobau. Or indeed might have halted Zieten's corps, hastening to Wellington's left flank.

Although Grouchy outnumbered Thielemann two to one (33,400 to 15,200), he did not fully appreciate it. In the meantime, Vandamme, without orders, moved to attack the Prussians. General Habert's 10th Division cleared Aisemont then attacked the bridge at Wavre, resulting in the death of Habert and 600 of his men. The Prussians were forced out of the suburbs on the right bank, but the French could not get over the river. Two further assaults were halted in their tracks. The 10th Division found itself pinned down along a 2-mile front by heavy Prussian musket and artillery fire.

Only after two hours of fighting did Lieutenant General Gérard's 4th Corps appear to renew the attack. On his arrival, Grouchy decided to launch two flanking attacks at Basse-Wavre and Bierges. He was still awaiting Pajol's cavalry. Hulot's division, which had been sent south of Wavre to take the crossing over the Dyle at Bierges ended in disaster.

Frustrated at every turn, Grouchy and Gérard led an unsuccessful assault on the bridge at the Mill of Bierges to the south of the town. With Gérard, Grouchy then led another attack.

Grouchy then changed tack and moved on Limale, 2 miles south-west of Wavre; the timely capture of this bridge could have allowed him to march directly on Waterloo. He received the warning of Bülow's attack on Napoleon's right at 5 p.m., it was much too late in the day to march on Mont-St-Jean and he was reluctant to move any of his forces westward towards Napoleon if he was to prosecute his attack on Wavre. Even at this stage though, Exelmans and Pajol might have served to confirm Gneisenau's fears. Leaving Vandamme and Exelmans to prosecute the attack on Wavre, and Hulot's division at Bierges, Grouchy arrived at Limale to find Pajol's cavalry ready for action.

In the meantime, Wellington desperately needed aid on his north-eastern flank as well as eastern. It was Gneisenau who had advised Blücher to send Zieten's 1st Corps directly to help Wellington, while the 4th was used to attack Napoleon's right rear with a view to cutting the Charleroi road, trapping the French Army. Zieten was in fact marching away from Wellington, convinced that the duke was defeated and having been ordered to help at Plancenoit. Fortunately, General Müffling, Wellington's Prussian liaison officer, managed to persuade Zieten to retrace his steps to come to the timely aid of Wellington's left wing.

In light of Napoleon's message, Grouchy could not disengage Vandamme, but ordered some of his forces to move on Limale. His plan was to attack St Lambert with half his force catching Bülow, while Vandamme dealt with what he thought was the rearguard of one Prussian brigade. The unobstructed bridge at Limale was swiftly taken by Pajol's hussars. General Teste's Division from 6th Corps supported them and occupied the nearby heights. At last, the road to Mont-St-Jean lay open, but it was too late.

In response, Thielemann moved his 12th Brigade from Bierges to Limale. The brigade launched a night attack but failed to drive off the French. At Wavre, fighting continued almost until midnight as 3rd Corps' thirteen attacks were driven off by four determined Prussian battalions. On five occasions the defenders drove them back across the river. Grouchy thought he had trapped Bülow, whereas Thielemann's outnumbered Prussians had held on covering Blücher's transfer of 72,000 men to aid Wellington.

At Waterloo, the French attacked the Papelotte–La Haye area again at about 6.30 p.m. Zieten's Prussians came to their allies' assistance sometime after 7 p.m., but once again the issue of uniforms caused problems. Saxe-Weimar mistook them for Grouchy and the Prussians thought the Nassau troops were French. Before the mistake was realised both sides set about each other with some vigour. The Prussians suffered heavy casualties before forcing the Nassauers from their barricades. Only then did the fighting stop. At Plancenoit, Bülow, reinforced by Pirch's 2nd 4Corps, retook the village only to be ejected by elements of Napoleon's Old Guard supporting the Young.

By this stage, Wellington's beleaguered and battered army had shrunk by about 50 per cent. Napoleon, in an act of final desperation, threw the Middle Guard against the duke at about 7 p.m. Two columns of Imperial Guard, the first numbering 4,000 men, plodded forward. Cruelly Napoleon spread the rumour that Grouchy had arrived on the field to stiffen their resolve, but they were halted in their tracks by musketry and swept away when Wellington ordered a general advance. The cry of treachery went up as it was realised that the distant advancing Frenchmen were in fact Prussians. The exhausted main French Army collapsed into chaos. Although the Prussians broke through on the French right at Plancenoit, Lobau and his brave men managed to escape encirclement.

To the west, during the night, with occasional fighting the French 4th Corps aided by Pajol expanded its bridgehead at Limale. Blissfully unaware of events at Waterloo, Grouchy futilely prepared to renew the battle on 19 June. Thielemann learned of the outcome at Waterloo and assumed Grouchy would withdraw at daybreak. He thought that those troops still facing him were a rearguard and as a result some Prussian troops were drawn off.

Instead Grouchy prepared to launch twenty-eight French battalions against ten Prussian and in the morning Gérard, Teste and Pajol were thrown into the fray. Upon hearing confirmation of the allies' victory, the Prussians renewed their own attack but were outnumbered two to one and were driven back. The French pressed home their assault and Bierges fell to Teste at 9 a.m.

Thielemann, with his centre pierced and unable to sustain his wings, decided not to continue the pointless battle. He had done what was required of him. He had succeeded in holding Grouchy at Wavre and drawn him well over the Dyle at Limale. Even the fighting on 19 June had not been a waste for the Prussians, for it had given Pirch's 2nd Corps time to block Grouchy's line of retreat. Exhausted, Thielemann began to retire, having endured ten hours of bitter fighting. Then, at 10.30 a.m. Grouchy learned of the disaster that had befallen Napoleon at Waterloo and ordered his own withdrawal, although he did not know the extent of Napoleon's defeat.

The Prussians suffered some 2,500 casualties at Wavre and a further 7,000 at Waterloo. Grouchy lost some 2,600 men at Wavre, while at Waterloo Napoleon suffered about 25,000 casualties, as well as 8,000 prisoners. Wellington's Anglo-Dutch losses at Waterloo were approximately 15,000.

Nevertheless, Grouchy still had some 30,000 formed troops under his command and the fighting was far from over. Despite defeat, General Vandamme showed a flash of inspiration. He daringly suggested marching on Brussels via Wavre, spreading panic and alarm, to force the allies in pursuit of Napoleon to turn back. They could then withdraw to France. Blinkered, Grouchy rejected the idea. Only now did he show any spirit in extricating his men. He escaped the Prussian 2nd Corps, and on 20 June beat off Thielemann's vanguard near Namur. The French 21st Division bravely stayed in Namur to cover the retreat. Grouchy crossed the Sambre with the Prussians close on his heels.

Napoleon, at this stage, had not given up the fight. He ordered France's fortresses not to surrender to the allies, who were advancing on Paris in parallel lines, and ordered the field army, which had lost half its strength in four days, to converge on Paris. He still had 117,000 men available to defend the city. Wellington established his headquarters at Malplaquet on 21 June, the scene of Marlborough's famous victory in 1709. The same day, Napoleon reached Paris but found the Chamber of Deputies would not support him. He was reluctant to stage a coup and was forced eventually to abdicate.

The Prussians, on reaching Senlis, were attacked on 27 June by Lieutenant General Kellermann's 1st Cuirassier Brigade. Also, D'Erlon's 4,000 men, who had rallied after Waterloo, came into contact with the Prussian 4th Corps. At Villers-Cotterêts the French were surprised and Grouchy was almost captured. He rallied 9,000 men and was reinforced by Vandamme, causing the Prussians to fall back. On the same day, the Anglo-Dutch Army crossed the Somme, while on 28 June the Prussians cut off the French retreat along the Soissons road.

Grouchy moved to protect northern Paris with the 1st, 2nd and 6th Corps, while Vandamme moved south with the 3rd and 4th. On paper the defence of Paris looked promising. Marshal Louis Nicolas Davout, minister of war, had some 60,000–70,000 men, as well as some 30,000 Paris National Guard with 300 guns. Napoleon offered to command them, but was rejected by a war weary administration. He departed on 29 June 1815 and HMS *Bellerophon* captured him attempting to escape to America. In the meantime, the Prussians drove the French from Aubervilliers. The French retook it but were again driven out, this time by elements of the British 4th Division, in what was Britain's last action in continental Europe until 1914. Elsewhere, the French *Armée du Rhin* was defeated at Strasbourg on 29 June by the Württembergers.

In the aftermath of Waterloo, France's political will to resist invasion was broken. Despite defeating the Prussians north of Versailles, the French military agreed to evacuate Paris and the triumphant allies entered on 7 July. By the end of the month nearly 700,000 allied troops had moved into France, making any further French resistance completely futile.

Waterloo has gone down in the lexicon of classic battles with Ligny and Quatre Bras as its precursor. Nonetheless, Wavre proved to be the critical battle, and for a crucial forty-eight hours Thielemann's corps had tied up the French right wing. Thielemann successfully prevented two much-needed French corps from reaching Napoleon. He had also protected the rear of Blücher's three vital Prussian corps that had gone to the assistance of Wellington. Grouchy, failing to realise that the Prussian Army had completely slipped his grasp, never appreciated the gravity of the French predicament at Waterloo.

Where Ney had thwarted Napoleon's decisive victory over the Prussians at Ligny, so too had Grouchy thwarted Napoleon's efforts at Waterloo. In the grand scheme of things, Ligny and Quatre Bras mattered little, the protagonists were like prizefighters squaring up. In contrast, Wavre proved to be the lynchpin, for Napoleon got it wrong on a day when he could ill afford to do so. Grouchy compounded this error by being the good soldier – when in doubt, follow orders. At Marengo in 1800, Marshal Desaix saved the day – at Waterloo, Grouchy did not.

Waterloo is full of endless what-ifs, especially with regard to Wavre. Gérard moving to the sound of the guns may have seen a different outcome; if not a French victory, then possibly all parties withdrawing to fight another day. Could Exelmans and Pajol have caused Zieten to retire leaving Wellington to buckle? None of this really matters, however. The fact remains that Napoleon lost largely through his own stupidity, even though the Prussians gave him a golden opportunity by arriving at Waterloo so late in the day. He brought about his own demise, for it was he who committed the fatal error of dividing his army at the deciding moment. He promoted Grouchy above his abilities, he appointed Grouchy as commander of the right wing of the French Army and he rejected Soult's advice to recall Grouchy earlier.

Despite his successful retreat on Paris, Grouchy was forever cast as the villain of the piece. He lived in exile in America until 1820 when he was allowed to return to France. Even so, his actions at Wavre forever haunted him.

12

Spear v. Rifle: Isandhlwana 1879

The Anglo-Zulu War of 1879 was the epitome of the colonial conflicts fought during the eighteenth and nineteenth centuries in which massed native armies were pitched against small European forces equipped with modern superior firepower. In the case of Isandhlwana, British complacency ensured that things did not go according to plan.

The war was a classic example of a pre-emptive strike and resulted from three principal factors: the considerable military power of the Zulu nation, the policy of the South African colonies and the character of the Governor of Cape Colony, Sir Henry Bartle Edward Frere. The overriding fear was of the Zulus' 40,000 strong *impi*, or army. The whole Zulu way of life revolved around military service, and the British were convinced that Zulu males were only granted permission to marry after the 'washing of the spears'. The implications of this were only too clear to the British administration in the Cape.

Britain had acquired South Africa, or rather the Cape, for naval and mercantile strategic reasons during the Napoleonic

Wars. The imposition of the British way of life on the exist-
ing predominantly Dutch settlers led to their mass migration
northwards, to the areas that were to become the Orange Free
State, Natal and the South African Republic (Transvaal). One
of the direct causes of the Dutch farmers' (Boers') treks north
was the abolition of slavery by Britain. It was felt that Britain
was overtly pro-native. The Boers harboured an understandable mistrust of the native populations, especially after 1838
with the betrayal and murder of Piet Retief and his followers
at the hands of the Zulu. Consequently, 10,000 Zulu were
defeated at the Battle of Blood River.

Conflict occurred in 1842 when the British authorities in
Cape Colony collided with the Boers during the Congella
incident near Durban in Natal. After a clash of arms, Britain
formally annexed Natal the following year, and set the stage
for the Anglo-Zulu War and the Transvaal (or First Boer)
War of 1881. In the meantime, the establishment of the
South African Republic and the Orange Free State by the
defiant Boers was recognised by Britain in the Sand River
Convention of 1852.

It was not until the 1870s that British policy began to centre
round the idea of a South African Confederation, under
British suzerainty. This would ensure British political and
economic dominance of the whole of South Africa. In 1871,
the Transvaal was denied economic security, when Britain
took over the diamond fields on the banks of the Orange
River and the Lower Vaal. Also, as a result of bad administration the Transvaal was teetering on the brink of bankruptcy.
In addition, there was a danger of war between Transvaal and
Zululand, which was in turn a threat to the security of Natal
and Cape Colony.

To Great Britain, the potential annexation of the Transvaal
provided an important opportunity; it was looked upon by

the colonial secretary, Lord Carnarvon, as a significant step towards establishing a loyal self-governing federation of the South African colonies that would take responsibility for its own defence. By annexing the Transvaal, it was hoped the Orange Free State would join voluntarily. Ironically, it was also hoped that British strategic and economic interests would be safeguarded, and involvement in South Africa lessened, with a reduced cost to the British taxpayer.

Invited in by a minority of Boers, the Transvaal was secured without a drop of blood being spilt. But the impact and implications of the annexation were to have far-reaching effects. Firstly, Britain had to bear the full brunt of the ensuing Zulu War which, initially aggravated by the Boers, was made inevitable by the attitude of the British administration. Significantly, hardly any Boers supported the British forces in a campaign that very publicly illustrated all the British Army's shortcomings.

Governor Frere saw war between the whites and Zulu as inevitable, and a string of boundary disputes sealed this view. The main incident in July 1878 involved a group of Zulu pursuing some Zulu women over the Natal border, capturing them, returning over the border and executing them. The white colonists of Natal would not tolerate this sort of arbitrary justice. Then, on 11 December 1878, the Boundary Commission, attempting to define a Natal–Zululand border, came down in favour of the Zulu instead of the Natal Boers. The Zulu, however, were also given an ultimatum that included the disbanding of their entire army. The Zulu King Cetshwayo could not accept all the demands, nor did he have time to, war was imminent.

Exactly a month later on 11 January 1879, Lord Chelmsford, Lieutenant General Frederic Augustus Thesiger, commander-in-chief of the British forces in South Africa, invaded Zululand leading a punitive expedition and war was declared. By this stage, the British Army was armed with the lever action breech-loading Martini–Henry rifle. This was vastly superior to the Brown Bess musket used throughout the Napoleonic Wars and was sighted up to 1,000 yards. It fired a hardened lead bullet, with a velocity of 1,350ft per second, that smashed bone and muscle. The muzzle could be fitted with a socket bayonet for close-quarter fighting. The cavalry were also issued with the shorter Martini–Henry carbine. The simplicity of operation and swift reloading resulted in the rifle being viewed as the first real successor to the longbow of the Hundred Years War. Before it, every firearm had been inferior to the master of Crécy, Poitiers and Agincourt.

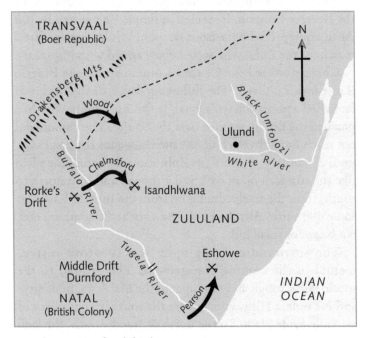

British invasion of Zululand, 1879.

The Martini–Henry, however, had a nasty habit of jamming and fouling, its recoil resulting in a vicious kick.

Chelmsford's plan was relatively simple – in order to counter any attempted Zulu incursions into Natal, he divided his forces (some 17,900, of which 9,350 were black auxiliaries) into five columns. Three would be placed along the Zulu border, with a fourth held at the Middle Drift on the Tugela River. The three main columns would then converge on the Zulu capital, Ulundi, while the fourth and fifth remained a buffer against potential Zulu invasion. The Northern Column was placed under Colonel Evelyn Wood, the Eastern under Colonel Charles Pearson, while Chelmsford accompanied Colonel Richard Glyn with the Central Column. Brevet Colonel Anthony Durnford was given command of the Reserve Column. It seemed so simple – a modern British Army equipped with the latest weapons would be more than a match for the Zulu, who were largely armed with just spears.

Chelmsford led the Central Column across the Buffalo River into Zululand. The following day, a small attack was launched against Sihayo's *kraal*, it was his people who had pursued the fleeing women over the border. The column did not reach Isandhlwana Hill, the site chosen for the first camp, until 20 January. The hill (its Zulu name denoting 'the place like the stomach of an ox'), is flat topped and runs north to south, from the track coming up from the Buffalo River for some 400 yards. Attached to it, via a *nek* at the southern end, is a *koppie*, or small hill.

The surrounding geography was to prove instrumental in the coming engagement. The ground to the south-east is rough and rises up into the Malakata, Inhlazatye and Nkandhla Hills, which form the southern boundary of a 4-mile-wide plain. To the north of Isandhlwana, the land climbs along a spur for 1,500 yards and links with the Nqutu Plateau, which stretches north and east for 10 miles to meet the Isipezi Hill. This forms the effective northern boundary of the plain. The main features of the plain are a large *donga*, or river bed, which lies at right angles to Isandhlwana Hill, while to the right of the *donga* lies a conical *koppie*.

The column arrived at noon and proceeded to establish camp on the eastern side of Isandhlwana Hill. The benefits of the site were that wood and water were readily available. Also, the Zulu Army marching from Ulundi in the east would have to cross the coverless plain. The camp was formed in a north–south line consisting of: the 2nd/3rd and 1st/3rd Natal Native Contingent (NNC), 2nd/24th Foot, N Battery, 5th Brigade Royal Artillery (RA), the Mounted Volunteers, and then the 1st/24th Foot south of the track. The tents ran for 800 yards and in between each unit were sited their wagons, although the main transport park was on the *nek*.

The first faults in the camp's security quickly emerged. Colonel Glyn wanted the wagons *laagered* into a defensive circle, but Chelmsford refused because it was impractical. First, it would take too long and exhaust the already tired oxen. Second, many wagons were to return to Rorke's Drift for supplies. Third, no entrenching was carried out because the ground was too hard and there were too few pioneers. All this was contrary to regulations issued to the Field Force in November 1878, which laid emphasis on defence, stating, 'When halting … form a wagon *laager*'. The Boers, who had defeated the Zulu in this way, forewarned Chelmsford, but he was unaware of the speed an *impi* could move; he saw a *laager* as a continual encumbrance that would slow the advance on Ulundi. The Boers knew differently from bitter experience, the only way to effectively fight the Zulu was from behind the sanctuary of a tight defensive wagon circle.

Many officers, including Lieutenant Melvill of the 1st/24th, displayed dissatisfaction. He complained, 'These Zulus will charge home and with our small numbers we ought to be in *laager* …' However, nothing was done, Chelmsford was sure he could win a face-to-face engagement and his pickets would give him all the warning needed. Foot pickets were posted in a half-mile arc, but in general the mounted vedettes were not posted well and there were none west of Isandhlwana, or on it.

Chelmsford and many of his senior officers could not conceive that the Zulu would dare to attack them in the open. Rather, they feared the Zulus would fight a guerrilla war similar to the Xhosa, who had fought nine Cape Frontier Wars. In his favour, Lord Chelmsford wanted to ensure the immediate area was free of Zulus before he continued his march on Ulundi.

A reconnaissance was led by him eastwards for 10 miles to the Nkandhla Hills. The Mangeni Valley and the empty *kraals* of Matyanna were surveyed from a ridge. Chelmsford had been informed that the *impi* had left Ulundi on 17 January, and with a distance of some 60 miles to cover it could be expected sometime after the 20th. It seemed logical that the *impi* would shelter in the southern hills, but with no sign of any Zulus, Chelmsford was back in the camp by 6 p.m. A Zulu was caught on the Nqutu and stated significantly to his captors, 'Why are you looking for the Zulus this way? The big *impi* is coming from that direction', and with that, he pointed to the north-east. Chelmsford though, disregarded this vital intelligence.

Having scouted the south-east, Chelmsford wanted the east scoured again before he pressed on. This would be carried out on 21 January and in two parts: Major Dartnell, with 150 Mounted Volunteers, would head south-east, and Commandant Lonsdale, with two battalions of the 3rd NNC, would go south round the Inhlazatye to link up with Dartnell. Chelmsford's reasoning was that the *impi* must be somewhere near the Isipezi Hill, they would not wish to cross the coverless plain and would either head north for the Nqutu or south to the Inhlazatye. Dartnell was searching the south and he would check the north. At 3 p.m. Chelmsford marched 3 miles across the Nqutu, but all he saw were fourteen mounted Zulus.

Meanwhile, Dartnell, 10 miles from the camp, ran into some Zulus moving north-east, he sent a message back for aid. Chelmsford was faced by Dartnell's message on returning from the Nqutu. The last thing he wanted was his army divided but it was too late to recall Dartnell, so supplies and permission to attack were sent. Dartnell now became aware that he was facing some 1,500–2,000 Zulus. The NNC formed a square and settled down for an uneasy night. Another messenger was sent and reached Chelmsford at 1.30 a.m.

Chelmsford was sure the *impi* was not in the Nqutu and therefore concluded that it must be near the Isipezi Hill to the south. If that was so, it meant Dartnell was facing the Zulu vanguard – he and Lonsdale would be in grave danger and must be reinforced. Chelmsford was, in fact, correct. On 20–21 January the *impi* had camped off the Ulundi track, north-west of the Isipezi Hill. It was just 15 miles from the camp and 8 miles from the force Dartnell had located. The Zulus, knowing the British camp had been weakened by Dartnell's departure, moved on the 21st onto the Nqutu, 3 miles from Isandhlwana. They planned to attack on 23 January 1879.

On the morning of the 22nd, Chelmsford summoned Colonel Glyn and Lieutenant Colonel Crealock in order to issue two very important orders: first, Brevet Lieutenant Colonel Pulliene, who was to remain behind, was to fight a purely defensive action if the need should arise; second, Durnford, with his 300 mounted natives, three companies of NNC and Major Russell's Rocket Battery, was to come up from Rorke's Drift and reinforce the camp. The order simply stated, 'You are to march to this camp at once with all the force you have with you of No. 2 Column ... 2/24th, artillery and mounted men with the General and Colonel Glyn move off at once to attack a Zulu force about ten miles distant.' Nowhere did it stipulate that Durnford was to take command at Isandhlwana.

At 2 a.m., reveille was sounded and the relief force for Dartnell was organised. It was to consist of six companies of the 2nd/24th, eighty-four mounted infantry, Colonel Harness' four guns of N/5 RA and the company of Natal Pioneers. The column moved off at about 3.30 a.m. and reached Dartnell at about 6 a.m., only to find the Zulus had gone. Their combined forces now totalled 2,500 men, 800 of whom were Europeans.

Chelmsford decided to scour east towards the Mangeni River. Lonsdale was to head north along the Mangeni Ridge, while Dartnell pushed round to the south. Glyn and the column plodded slowly across the plain towards Lonsdale, who skirmished with some Zulus holed up in caves, killing thirty. By 9.15 a.m. though, all the Zulus had vanished eastwards, and Glyn arrived at 9.30 a.m. Seeing that the *impi* was not in his immediate area, Chelmsford decided to set up camp in the Mangeni Valley, which Pulliene could be brought up to. Commandant Hamilton-Browne, with his 1st/3rd NNC were ordered to return to the camp to collect rations and tentage.

Just after Glyn arrived, a messenger from the camp rode in bearing a note, it said, 'Report just come in that the Zulus are advancing in force from left front [Nqutu] of camp. 8.05 a.m.' Chelmsford was not perturbed by this, and he could not be back before 12.30 p.m., so he decided to carry on with his original plan. Captain Gardner was sent to proceed Hamilton-Browne with orders to pack up.

Chelmsford also sent Lieutenant Milne, his naval ADC who possessed a telescope, to a nearby rise to observe the camp. The camp at Isandhlwana was 12 miles away, but as far as he could discern there was nothing to worry about. At 10.30 a.m. Chelmsford decided to reconnoitre the Mangeni Valley.

When Chelmsford left Isandhlwana, Brevet Lieutenant Colonel Henry Pulliene was left in command. Under him were 1,800 men, 950 of whom were Europeans. The troops allocated to defend the 900-yard frontage were: five companies of the 1st/24th, one company of the 2nd/24th, 165 Mounted Volunteers, seventy-two artillery-men with two 7-pounders, and four companies of the NNC. They were dispersed as follows: two companies of NNC to the left of the camp and one on the spur leading up onto the Nqutu. Towards the big *donga* in front of the camp was G Company 2nd/24th under Lieutenant Pope; this unit was stretched out on a 1,000-yard frontage.

At 8 a.m. a rider arrived from the Nqutu. He informed Pulliene that a large body of Zulus were crossing the plateau from the north-east. The 'stand to' was sounded and a messenger sent to Chelmsford, who reached him at 9.30 a.m. The troops stood about for an hour, but all that was seen were a few Zulus 3 miles away on the escarpment. It was not until 9 a.m. that a second rider arrived from the Nqutu, to say the Zulus had split into three bodies, two had moved north-east and vanished, the other to the north-west.

Colonel Durnford rode in at 10 a.m. He should have technically taken command, because he was three years senior to Pulliene. Although there was no specific order placing Durnford in overall command, Pulliene made it clear that he did not mind handing over his command. After breakfast, yet another trooper arrived from the Nqutu, but he was so winded his message was quite incoherent.

Durnford decided to send Lieutenant Roberts with his troop of Sikali Horse to scout the north-west of the plateau, while Lieutenant Raw headed north-east. Lieutenant Barry's company of NNC were to support the horse, while Captain Cavaye's E Company 1st/24th moved onto the spur to take Barry's place. Meanwhile, Hamilton-Browne was about 9 miles from the camp when he caught two Zulus. Alarmingly, they informed him the *impi* was in the north, poised to attack the camp. A messenger was sent to warn Chelmsford, but he did not reach him until 2 p.m.

Durnford and Pulliene were informed that about 500 Zulus had been spotted to the north-east moving east. To Durnford, it appeared that they must be moving against Chelmsford's rear, in order to block his communications with the camp. Durnford decided they must be stopped. With his last remaining mounted troops, Lieutenant Henderson's Sikalis and Lieutenant Newnham-Davis with the Edendale contingent, he planned to trap or block the Zulus in between Raw and Roberts. They were to move along the plateau and link up with him in the east. Durnford also wanted two companies of the 24th, but Pulliene politely refused, no doubt with Chelmsford's orders on defensive action still in mind. At 11.30 a.m., Durnford rode out with the two troops of horse, D Company NNC and the Rocket Battery. He soon passed the conical *koppie* heading north-eastwards, but the Rocket Battery and NNC lagged behind and veered further north.

Up on the plateau, Roberts moved along the escarpment and Raw spread out to the north-east. Suddenly, some of the furthest away troopers spotted a handful of Zulus and cattle, 4 miles north-east of the spur. With a cry, the Basuto scouts sped after them, but they vanished over a ridge. A solitary trooper rode ahead, only to rein up hard at the edge of a huge ravine. He stared down at the floor and sides of the ravine – they stretched as far as the eye could see, a mass of some 20,000–25,000 Zulu warriors. The man did not know, but a mile behind this ravine lay another with the late arrivals; the Undi Corps and uDloko Regiment, they numbered another 6,000 plus. The trooper gaped, here lay the *impi* only 5½ miles north-east of the camp, some 11 miles north of where Chelmsford was fruitlessly searching. The rider turned and fled towards Raw.

The *impi* had been awaiting the following day to attack, but waiting was now out of the question. The Zulus had been squatting silently, but they now began to rise – first hundreds, then thousands. The mass began to seethe like a great wave, building momentum the warriors poured up the slope and over the edge. The first regiments out were the umCijo, umHlanga and a few uThulwana. The Zulu left horn, consisting of the inGobamakhosi, umBonambi and the uVe, began to fan out across the plain, racing to close with the right. The right horn, consisting of the uDududu, Nokenke and Nodwengu, raced across the British front in order to move behind Isandhlwana Hill to seal the *nek*. The chest, or centre, of their army was formed by the umKhulutshane, umHlanga, isAnqu and umCijo.

The Zulus wanted to close with their enemy quickly, for they excelled in hand-to-hand combat and knew the devastation that firearms could inflict on their ranks. Their primary weapons were the throwing and stabbing spear. It has been estimated that the Zulus had up to 20,000 guns, of which only 500 were modern breech-loaders, 7,500 were percussion weapons and the rest obsolete muskets. Some of them were up to 40 years old. Furthermore, the Zulu on the whole were not good marksmen.

Raw, supported by Roberts, hastily spread his men out in order to gain the camp time. Their 100 carbines fired again

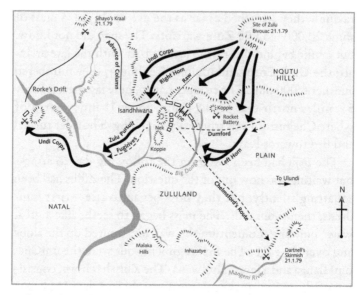

Zulu *impi*'s attack on Isandhlwana, 22 January 1879.

and again but had absolutely no effect, so the two troops covering each other began to retire. Barry's NNC fled at the sight of the oncoming Zulu horde. Shepstone, from Raw's troop, sped down the spur to warn Pulliene, while another rider headed for Durnford. He was 4 miles east of the camp when he spotted the Zulus, 1,500 yards away pouring down the escarpment. Firing as they went, his troopers began to slowly retire. On his left, Russell and Nourse's NNC bravely prepared to face the Zulus.

The camp now contained the following companies of the 1st/24th: Lieutenant Porteous' A, Captain Younghusband's C and Captain Wardell's H, plus Pope's G Company, of the 2nd/24th, two 7-pounders, two companies of NNC, thirty Mounted Volunteers and Captain Lonsdale's NNC just to the north-east. The Zulus were now only 2 miles to the north-east and approaching fast.

Worryingly, Pulliene found he could not fight a compact defensive action because Durnford, who was fighting a rear-guard, was in severe danger of being cut off. If the Zulus reached the big *donga* before Durnford passed it, he would be trapped. Therefore, Pulliene could not contract his frontage. Instead, he disastrously proceeded to extend it. Captain Mostyn with F Company 1st/24th was sent up the spur to support Cavaye. The frightened NNC on the spur began to retire, while Younghusband with C Company was sent to the north on the left, to cover the retiring companies.

The three companies facing east were stretched on a frontage of 3½ miles: to the south was Lieutenant Pope's G Company 2nd/24th, then Captain Wardell's H Company and Lieutenant Porteous' A Company. Each company was separated by a gap of some 200 yards, while the north-eastern corner, between Cavaye and Porteous, had a gap of 300 yards. Major Smith's guns were placed in this gap.

Most accounts place two companies of NNC in between Cavaye and Porteous as well, but it seems ridiculous to believe that regular troops would rely on levies to hold such a key point. One answer is that some of the fleeing NNC from the plateau stopped in the gap for a short time. It seems more likely that the regular companies edged towards each other, stretching their spacing more than ever.

Durnford had unfortunately ridden past the conical *koppie* before he saw, to his horror, the entire Zulu *impi* appear on the edge of the plateau, stretching for some 2 miles. He now saw the danger that Pulliene had foreseen, and immediately began to quickly retire, leaving Russell still frantically setting up his rockets!

The Zulus swarmed down off the plateau, having run about 5 miles, and they began to near the British line. To the north the Nokenke, numbering some 2,000 warriors, rushed down the spur, only to be cut to pieces by the well-ordered volleys of C, E and F Companies. Blasted beyond endurance they began to retire back up the spur. The isAnqu regiment, about 1,500 strong to the Nokenkes' right, were also halted at 300 yards. However, Smith's guns, firing case shot, were soon found to be having little effect, because when the gun crews stood clear to fire their artillery, the Zulus cried, '*Umoya!*' (air) and threw themselves to the ground. The Zulu centre, numbering about 10,000 and consisting of the umCijo and umHlanga regiments were halted at about 400 yards. But Pope, Wardell and Porteous were holding a frontage double of that covered by the three northern companies and the Zulu began to edge forward slowly.

On the right the 6,000 inGobamakhosi surged forward and overran Russell, killing six and wounding three of his crew. They also broke D Company of the NNC. Durnford was still retiring and now saw his left flank exposed. Ordering his troops further back, they hurriedly lined the big *donga* running parallel with the camp. Durnford saw that he dared not retire because the Zulu left horn would outflank the camp. He was speedily reinforced by the Newcastle Mounted Rifles who, along with his troops, began to hold the Zulus at bay. The inGobamakhosi converged with the 4,000 uVe, but they could not get near the *donga*. Then, one of Smith's guns was brought up and it dispersed them, but the 2,000 umBonambi began to move to their left in order to outflank Durnford's right wing.

For all the Zulus' efforts, they could not get near the British positions in order to make use of their superior numbers and close-quarter fighting methods they had been taught since their teens. The great mass of men lapped round the British line, and everywhere lay Zulu dead and wounded. To the British troops it was obvious they were giving the 'savage' a good beating. The battle stabilised for about fifteen

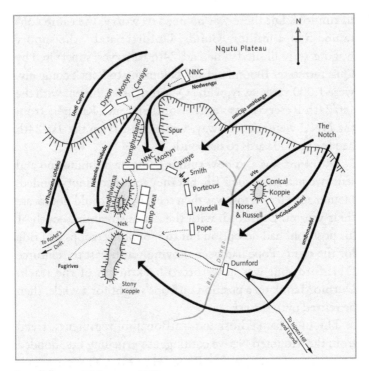

Isandhlwana, 22 January 1870.

minutes, the Zulus receding here and there, crawling forward or futilely sheltering behind their shields.

Pope, seeing the danger to his right, had wheeled about to face east and link up with Wardell. Mostyn and Cavaye edged right, in order to link with Porteous. This left a gap between Younghusband and Mostyn, but some NNC moved to fill it. By 1.30 p.m. the British were in a complete arc, and the *impi* was halted at about 400 yards everywhere. It seemed that European firepower was winning the day.

Cavaye had now been firing for an hour and the other two companies for half an hour. Ammunition was beginning to run low, but there was no need to worry, the camp contained half a million rounds. Unfortunately, the supply wagons were ill sited. The 2nd/24th Reserves, supervised by Quartermaster Bloomfield and belonging to Pope's company, were 1,000 yards away, while Quartermaster Pullen, with the 1st/24th Reserves, was at the southern end. Runners from the NNC were turned away, while those from the 1st/24th had to run 500 yards to the south.

Durnford was also now running low on ammunition and sent two natives to the 24th, but they returned emptyhanded. Lieutenant Henderson was then sent, but he could not locate their wagon. Durnford, with the Zulus streaming south of his position, had no option but to mount his troops and ride for the camp. Pope had been moving south-east to reinforce Durnford, but was now forced back north of the track. Durnford took up a position on Pope's right for a while, then he retired onto the *nek*.

The inGobamakhosi and umBonambi regiments, freed from the Mounted Native contingent's gruelling fire, flooded across the *donga* towards Pope. However, G Company's move, well intentioned as it was, now caused a 700-yard gap between itself and H Company. The whole of the right flank was exposed with only about ninety men to hold off thousands of maddened Zulu. If Lonsdale had any NNC in this gap, they had no doubt fled by now.

The centre had so far held the Zulus at about 150 yards, but with the continuing slackening of fire the Zulus began to stamp, bang their shields and cry '*Usuthu!*' (kill). Encouraged by their left, they swept forward. Any NNC on the corner had by now gone and Smith tried to withdraw his guns. As 'retire' sounded, the umCijo, Nokenke and uKhandempernvu regiments flooded through the 300-yard gap between Cavaye and Porteous. There followed a nightmare struggle as Smith's men fought to extricate their guns from the Zulus, and they only just succeeded. Ironically, Chelmsford was 10 miles away spying the camp with his field glasses and could see nothing wrong.

One Zulu veteran recalled that 'like a flame the whole Zulu force sprang to its feet and darted upon them'. They were in their element, and at last they could utilise their *assegais* (short stabbing spears). E and F Companies were attacked from behind and were hacked to pieces. With no time to retire, form square or even fix bayonets, they died to the man, including Mostyn and Cavaye. H and A Companies fared no better, pressed so close they could not fight off the thousands of stabbing Zulu. Wardell's troops had edged south towards Pope, and were so stretched out that the Zulus simply swept through them. Porteous, with A Company, suffered the same fate.

Pope's troops were facing the inGobamakhosi and umBonambi and were attacked in the rear, but they managed to make a fighting retreat up the track towards the *nek*. Younghusband, with C Company on the far left, had a brief breathing space during which they fixed bayonets. They then began to retire up the slopes of Isandhlwana; from there he beat off the Zulus, his sixty-odd men firing into the closely packed warriors.

Meanwhile Durnford, with about thirty-four mounted men, had joined Pope on the *nek*. From there, they frantically held back the Zulu left horn, while many fugitives fled behind them. To the north of the *nek*, some mounted troops also held off the encircling Zulu right. Durnford's volleys began to dwindle and finally the umBonambi pressed home. The carnage was terrible and Durnford's seventy or so men were completely overwhelmed.

To the north, Younghusband had finally run out of ammunition and his men were slowly forced off their rock platform into the milling warriors below. In the end Younghusband and his soldiers charged down the slopes, eventually dying in the wagon park. At 2 p.m. there was a partial eclipse of the sun and the Zulus set about finishing off the last pockets of resistance. Half an hour later there was only one white man left alive in the camp, lodged in the rocks, but he too was despatched when his ammunition ran out. The Battle of Isandhlwana was over.

All those still alive had by now fled over the *nek* towards Rorke's Drift. Unfortunately, the Undi Corps and the uThulwana regiment were blocking the track and the umBonambi were in full pursuit. Smith's guns only got half a mile from the *nek*, where they crashed over the edge of a gully. Lieutenants Melvill and Coghill, entrusted with the Queen's Colour of the 1st/24th, were both killed on the banks of the Buffalo River and the standard was lost in it.

Chelmsford did not decide to ride towards Isandhlwana in order to investigate until about 2.30 p.m. and it was not until 3.30 p.m., when he was 4½ miles from the camp that he became fully aware of the terrible disaster that had occurred. The camp had contained 950 Europeans and 550 natives, of which 858 Europeans and 471 natives were dead. Out of a total force of 1,800 men, 1,329 were dead. As for the Zulu, they suffered over 2,000 casualties. This was truly a costly victory.

Chelmsford's despatch, which informed a stunned London of the disaster, aptly sums up the battle:

I regret to inform a very disastrous engagement which took place on the morning of the 22nd January between the armies of the Zulu King Cetshwayo and our own No. 3 Column, consisting of five companies of the 1st Bat. 24th Foot and other ranks. The Zulus, in overwhelming numbers launched a highly disciplined attack on the slopes of the mountain of Isandhlwana, and in spite of gallant resistance, the column was completely annihilated.

The Zulus were successfully beaten off at subsequent engagements at Rorke's Drift, Eshowe, Hlobane, Kambula and Gingindlovu with the loss of about 3,000 warriors. Zululand was then invaded a second time.

Britain gained revenge for Isandhlwana at Ulundi, where 20,000 Zulus were defeated on 4 July 1879 with 1,000 dead. A British politician at the time put the total number of Zulu dead as high as 10,000. Their capital was burned and Cetshwayo eventually captured. The kingdom was divided up under rulers amenable to the British. Cetshwayo died in 1884 and Zululand was annexed to the British crown three years later. It was then annexed to Natal in 1897, becoming part of the very colony to which it had once posed such a threat. There were two brief rebellions in 1888 and 1906.

Just two years after the Anglo-Zulu War, Britain and the Boers fell out once more. The troublesome Boers who had stood by while Britain embarrassed itself were to have the last laugh. In the Transvaal War of 1881, British forces suffered three defeats at Laing's Nek, Ingogo and Majuba. The Battle of Majuba was a particularly humiliating engagement and was to be tragically mirrored by Spion Kop in 1900. Despite Britain's hapless response to the Boer invasion of Natal, both sides were eager to see the hostilities brought to a close.

The discovery of gold on the Rand in the Transvaal was a predominant factor contributing to the Second Boer War of 1899–1902. The conflict saw British forces again suffering badly during a transitional period in warfare. The initial Boer invasions resulted in a series of field engagements and sieges, such as Celenso, Spion Kop, Ladysmith and Mafeking.

At first the Boers fared well, but the sheer weight of British numbers forced them increasingly to adopt guerrilla tactics. Also, the British use of blockhouses and concentration camps gradually strangled the Boer cause, and by 1902, although not completely defeated, they were prepared to negotiate.

The British Government came under increasing internal pressure from public opinion and was also keen to bring the war to a conclusion. On 31 May 1902, in the Treaty of Vereeniging, the Boers formally surrendered their independence and Britain asserted her sovereignty, but this was to be a short-term victory. Britain agreed to the following: to pay £3 million for the rehabilitation of the two states; take no judicial proceedings against returning Boer soldiers; Afrikaans would still be taught; rifles could be retained on licence; and the military administration would be followed by a civil government that would be dominated by the Boers.

Although Chelmsford crushed the Zulu nation, his military career was ruined, for Britain never truly forgave him for the humiliating disaster that was the Battle of Isandhlwana. Bad luck, ineptitude and complacency characterised Isandhlwana. If Chelmsford had kept his command together and fought the Zulu by employing Boer tactics on the open plain, then the engagement may have gone very differently. Furthermore, Pulliene bears some responsibility. He should have impressed upon Durnford that their roles were to fight a defensive action, not take the initiative against the Zulu. In the end, No. 3 Column was just too spread out to ensure that its concentrated firepower would drive the Zulu *impi* off and it was overwhelmed.

13

TANK V. TANK:
VILLERS-BRETONNEUX 1918

The terrible global war that began in Europe in August 1914 became the bloodiest conflict in history. Despite the deadlock in the trenches, the First World War was characterised by great technological advances, especially with the birth of aerial and submarine warfare. German bombers and massive Zeppelin airships launched terror raids on British cities, while the U-boats almost strangled Britain's supply routes. Chemical warfare was unleashed by both sides with horrific results. Notably, the war also heralded the birth of tank warfare.

After the Napoleonic Wars, armies had fought using breech-loading single-shot rifles. By the 1900s, soldiers were issued with bolt-action rifles employing magazine-fed ammunition such as the British Lee–Enfield and the German Mauser. These offered a far greater rate of fire. The Gatling gun of the late 1800s evolved into the Maxim and Vickers machine guns, capable of spitting out 500 rounds a minute. The early machine guns were all hand cranked, but these used recoil to load, fire and eject continually while the trigger was pulled back. Ammunition was fed in by a flexible belt and the weapon was cooled by a water jacket round the barrel. Rapid-firing, breech-loading, long-range artillery also became the norm. All these weapons made advancing over open ground even more deadly. The result was that everyone dug in and developed vast trench networks.

By the spring of 1915, the fighting on the Western Front had reached stalemate. The initial war of mobility and manoeuvre had stagnated into static confrontation – artillery, machine guns and barbed wire became kings of the battlefield. Attempts by the British to breakthrough at Ypres failed and it was decided that brute force and attrition was the way to crush Germany. The German High Command concluded, after the use of poison gas had failed to achieve the desired results, that they would remain on the defensive while they concentrated on the Eastern Front and the war against Imperial Russia and her allies.

British boffins needed to find a solution to what had become effectively siege warfare on the Western Front. To break the deadlock, they turned to the one device that had dominated the first half of the twentieth century – the internal combustion engine, which ironically had been invented by the Germans Karl Benz and Gottlieb Daimler in 1885. Lorries and buses had already been pressed into service to ferry troops and supplies to the front but, while they were a great help, they struggled to cope with the seas of mud and vast craters that had developed around the trenches.

The very first British tank went into service in 1916, exactly two years after the outbreak of the war. Its rather peculiar and very distinctive rhomboid shape was designed to ensure a long track run, capable of crossing the vast trench networks on the Western Front. The tracks ran around the outer edges of the entire hull and could manage a very modest

3.7mph. Speed was not really an issue, however, as the tank was intended to match the pace of the infantry. The resulting height meant a turret would have made the vehicle top heavy, so the main armament was placed in sponsons on either side of the hull.

It was dubbed, rather unimaginatively, the Mark I, of which just 150 were built in two types, males with cannon and females with machine guns. The Mark I was blooded at the Battle of Flers–Courcelette, part of the renewed Somme Offensive, on 15 September 1916. British heavy tank designations ran from the Mark I through to the Mark X (although the VI, VII and X never went into production). The improved Mark II/III consisted of half-male and half-female variants. These were only intended as training tanks, but the Mark IIs ended up being shipped to France.

The key variant was the Mark IV, which was an uparmoured version of the Mark I, of which 1,220 were built, comprising 420 males, 595 females and 205 tank tenders. This tank weighed 28.4 tons and required an eight-man crew. The male was armed with two 6-pounder guns plus three .303 Lewis machine guns, while the female was equipped with five Lewis guns. On the Mark IV and V females, the sponsons were smaller, being only half-height top down. As with most British tanks of the time, it was slow, managing just 4mph depending on the conditions. During the summer of 1917, the Mark IV fought at Messines Ridge and the Third Battle of Ypres, then in November of that year 432 Mark IVs were committed to the Battle of Cambrai.

The smaller British Medium Mark A Whippet was designed to support the slow, heavy tanks by exploiting any breakthroughs. The first production tanks were delivered in October 1917, from an order for 200; this was increased to 385, but later cancelled. The Whippet was directly derived from the experimental 'Little Willie' prototype. While the Whippet looked more like the tank as we know it, the crew compartment consisted of a fixed turret at the rear of the vehicle. This was armed with four machine guns, but as there were only three crew it meant that the gunner had his work cut out. The Whippet, at 14 tons, could manage 8mph. This was Britain's answer to the Renault FT-17 Light, although it was double the weight and required an extra crewman.

At 6 a.m. on 20 November 1917 the British Tank Corps attack at Cambrai opened. Their commander, Brigadier Sir Hugh Elles, rode into battle in a Mark IV Male called 'Hilda'. The Mark IV was the first British tank to be used en masse and was the main type that fought at Cambrai. Apart from problems experienced by the 51st Highland Division, everything went pretty much to plan. Heralded by their roaring engines and great billowing clouds of petrol fumes, the massed ranks of 470 tanks crawled into no man's land and through the chaos caused by the opening bombardment.

The attrition rate for the Tank Corps was very considerable. By the end of 20 November 1917, a total of 179 tanks were reported out of action, with sixty-five destroyed, seventy-one broken down and forty-three ditched. However, this sacrifice had not been in vain. A breach 4,000 yards deep and 6 miles wide had been cut into the German defences at the cost of less than 4,000 casualties. In England, the church bells rang out at the news of the victory.

Supported by 100 tanks and 430 guns, the British assaulted the woods on Bourlon Ridge on 23 November but made little progress. Four days later the 62nd Division, aided by thirty tanks, renewed the attack. Once in control of the crest of the ridge, the British dug in to endure a deluge of 16,000 German artillery shells.

The Germans launched a counter-offensive at Cambrai at 7 a.m. on 30 November 1917 with the intention of retaking the Bourlon salient as well as attacking around Havrincourt. By the end of the Battle of Cambrai, the British had suffered 44,000 causalities for very little territorial gain. Nonetheless, British tank tactics had proved largely sound and had shaken the Germans. In London, leading politicians were far from happy about the outcome of Cambrai. 'This action was grossly bungled,' said Prime Minister Lloyd George, 'and the tank success was thrown away through the ineptitude of the High Command.' In his memoirs, Lloyd George was more specific, pointing the finger firmly at General Byng:

> The first onset of the Tanks, on 20th November, was a brilliant success. Within a few hours the Hindenburg line was broken by these inexorable machines, and a penetration effected in the enemy lines as deep as that which had been achieved after months of terrible fighting and colossal losses on the Somme and at Passchendaele. It is generally acknowledged now that the advance was badly muddled by General Byng and that he could, even with the resources at his command, have made a better job of it. But what converted victory into defeat was a total lack of reserves.

It was not an altogether auspicious start to massed armoured warfare. The Germans, however, finally began to take note of the possibilities offered by the tank which, to date, they had largely ignored. Ironically, while the Mark IV became the most numerically significant tank of the war, it also became the most prominent German one. They collected all the Mark IVs they captured at Cambrai and the earlier actions at Charleroi and refurbished them.

In German service, the Mark IV was dubbed the Beutepanzer Wagen IV or 'captured armoured vehicle 4'. In December 1917 the German Army formed four tank companies, employing around forty captured Mark IVs. However, the first tank-to-tank battle did not occur until April 1918, when cumbersome German-built A7Vs took on British Mark IVs at the Second Battle of Villers-Bretonneux. This heralded the true birth of tank warfare.

Thanks to the Allied naval blockade, the German population was suffering by 1918. German submarines had failed to starve Britain into submission or stop the shipping of American forces to the continent. Even by this stage of the war, the German High Command was not receptive to the tank. From their perspective, those deployed by the British and French had not achieved truly decisive results. They decided that better infantry tactics that would allow for greater initiative and individual firepower were needed. General Ludendorff, gathering his last reserves, gambled on a knockout blow against the British, who he considered the most vulnerable. If this could be achieved, the Allies might sue for peace and bring an end to the interminable conflict.

The German military felt that the anti-tank rifle, sub-machine gun and light machine gun offered much better prospects for achieving a breakthrough. Peace with Bolshevik Russia permitted Ludendorff to redeploy fifty battle-hardened divisions from the Eastern Front. Crucially, these included troops trained in the latest infiltration tactics. They employed a new type of soldier, dubbed the 'stormtrooper' and armed with the new Bergmann MP18 sub-machine gun. This greatly increased their firepower. They were trained to advance quickly through enemy lines leaving bunkers and other strongpoints to be mopped up by subsequent waves of infantry.

General Oskar von Hutier devised these new tactics (from which the Nazi concept of blitzkrieg, or lightning war, was to evolve) and they were first used on the Eastern Front at Riga in 1917. The Germans also produced the very first purpose-built anti-tank rifle called the Mauser T-Gewehr. It fired a 13mm round and was based on big game hunters' elephant guns. This was issued to specially formed anti-tank detachments whose job was to hunt down tanks.

Nonetheless, a rudimentary German tank programme was also grudgingly instigated. The first German tank, the unwieldy A7V Sturmpanzerwagen, went into production in October 1917. Although 100 were ordered, only twenty had armoured bodies as the rest were used as cargo carriers. With a crew of seventeen, the A7V was a lumbering armoured box armed with six 7.9mm machine guns and a 57mm Maxim-Nordenfelt cannon mounted at the front. It bore a greater resemblance to the French box-type heavy tanks than the British rhomboid design. This was quite surprising, as the Germans operated British tanks. Weighing some 33 tons when combat loaded, the A7V could manage a top cross-country speed of 4mph and a radius of about 20 miles.

In reality, the Sturmpanzerwagen was little more than an armoured gun platform for infantry support. As a result, the design suffered from all the same problems that plagued the unwieldy French-built Saint-Chamond and Schneider tanks. Its shape and size meant it was very poorly balanced and as the engine was underpowered it had very poor cross-country capability. A female variant of the A7V was designed with two machine guns replacing the main armament, but only one of these saw combat before being converted to take the 57mm gun. As Ludendorff only had a handful of A7Vs, they were not a factor in his Spring 1918 Offensive. Instead, he relied on the new infantry tactics to carry his troops forward to victory.

By March, Ludendorff had massed sixty-three divisions facing the 56-mile-long British sector between Arras and St-Quentin held by the British 5th and 3rd Armies. They mustered just twenty-six divisions – Gough's 5th Army was especially thinly spread in the south where it had been deployed to recuperate after its mauling at Passchendaele. Although better organised, Byng's 3rd Army, which lay to the north, would be exposed if 5th Army gave ground.

On 21 March 1918, some 9,000 artillery pieces and mortars heralded the German Michael Offensive. This was the opening stage of a whole series of attacks designed to crack open the Allied Western Front. General von Hutier's 18th Army set about Gough while General Georg von der Marwitz attacked Byng. The A7V proved mechanically unreliable and when five went into action that day north of the St-Quentin Canal, three immediately broke down.

Despite punching up to 40 miles in places, within a week the Germans began to lose momentum. By 28 March, General Hutier had come to a temporary halt and by 5 April, General Marwitz had been stopped by British and Australian troops at Villers-Bretonneux, some 10 miles short of his objective, Amiens. In two weeks of fighting, Ludendorff suffered 250,000 casualties, which included a large number of his specialist stormtroopers. Nonetheless, keeping the pressure up on 9 April, Ludendorff launched Operation Georgette to the north against Flanders with the aim of threatening the Channel ports.

The Allies reeled back in two areas under the shock of the German offensives. Ludendorff gained some ground south of Dunkirk and north of Arras. South of Arras, though, he achieved an even greater penetration extending to Reims. Allied tanks deployed in the defence were used as mobile pillboxes, which posed little problem for the highly mobile German stormtroopers. However, when the tanks were combined, they greatly helped in stemming the German tide.

If Ludendorff succeeded in taking Villers-Bretonneux plus the high ground between there and Cachy then German artillery would be able to dominate Amiens. Alert to the danger, the French moved forces including Foreign Legion units to Bois de Blangy, about 4 miles west of Villers-Bretonneux, along with A Company, 1st Tank Battalion of the British Tank Corps.

The tanks were deployed to conduct so-called 'Savage Rabbit' tactics, whereby they lay in ambush positions ready to surprise infiltrating enemy troops. A Company was equipped with patched up Mark IVs. No. 1 Section was commanded by Captain J.C. Brown MC, with three tanks – two female and one male. The single male was commanded by 2nd Lieutenant Frank Mitchell. Apart from its short 6-pounder guns and no steering tail, it was little different in appearance to the original Mark I. A second British tank type was involved in the coming battle, consisting of seven British Medium A Whippets from C Battalion Tank Corps, under Captain T.R. Price.

Captain Brown's tanks were given a warm reception when they moved into the woods just to the north of Cachy. On the night of 23 April, a German spotter plane flew over the Bois l'Abbé and dropped flares. These were followed by high-explosive rounds and mustard gas shells that forced No. 1 Section to withdraw to the western edge of the woods.

The Germans, by this stage, were short of troops and ammunition, which precluded anything more than a spoiling attack in the Villers-Bretonneux and Cachy area. However, they had fourteen A7V Sturmpanzerwagens available to support the operation. These were divided into three groups with the first comprising three tanks heading for Villers-Bretonneux, the second made up of seven tanks on the right flank heading for Bois d'Aquenne, and the third with four tanks driving on Cachy.

In the meantime, two of Lieutenant Mitchell's crew were overcome by gas fumes and had to be evacuated. Nearby wounded infantry informed Brown and Mitchell that enemy troops were in Villers-Bretonneux. It was decided that they would counter-attack under the covering fire of a local battery of 18-pounder guns. At 8.45 a.m. on 24 April 1918, Brown set off in Mitchell's tank, which deployed nearest the woods with the two female tanks on the right. 'As the wood was still thick with gas we wore our masks,' recalled Mitchell, 'while cranking up a third member of my crew collapsed and I had to leave him behind propped up against a tree trunk.' A crewman from one of the females was sent over to help, but Mitchell was still two crew short.

At 9.30 a.m., when they reached the defences of the Cachy Switch Line an infantryman warned them of the presence of German tanks. Mitchell saw three objects heading towards eastern Cachy followed by enemy infantry. The fourth German tank (called '*Elfriede*') with the group had got lost in the fog, veered north and fallen into a quarry. Captain Brown ran to warn the females while Mitchell swung right to move parallel to the nearest German tank under 2nd Lieutenant Wilhelm Biltz called '*Nixe*'.

Mitchell's left-hand 6-pounder began to fire on *Nixe* but there was no response, while his forward Lewis gun opened up on the German infantry. He then turned his tank to face the Germans. Mitchell later wrote:

Suddenly there was a noise like a storm of hail beating against our right wall and the tank became alive with

splinters. It was a broadside of armour-piercing bullets …
The crew lay flat on the floor. I ordered the driver to go
straight ahead and we gradually drew clear, but not before
our faces were splintered. Steel helmets protected our heads.

Lieutenant Biltz had put his tank into reverse and opened up
with everything he had. Mitchell, hoping to get a clear shot,
stopped. This action convinced Biltz that the British tank was
knocked out so he turned his attention on the two females.
He hit both, ripping holes in their armour and forcing them
to withdraw. In the meantime, Mitchell rumbled forward,
recalling at 10.20 a.m.:

> The gunner ranged steadily nearer and then I saw a shell
> burst high up on the forward part of the German tank. It
> was a direct hit. He obtained a second hit almost immedi-
> ately lower down on the side facing us and then a third in
> the same region. It was splendid shooting for a man whose
> eyes were swollen by gas and who was working his gun
> single-handed, owing to the shortage of crew.

Biltz's crew were in a bad way. Mitchell's first hit killed the
front gunner, mortally wounded two others and injured
a further three. Afraid that a box of hand grenades might
explode, the crew bailed out. However, Mitchell's other
two hits did little damage and *Nixe*'s engines continued running. Biltz and his crew returned to their tank and drove 2km
before the engines seized through lack of oil.

Mitchell engaged the other two A7Vs and when they
withdrew southwards he assumed he had seen them off. He
then continued to patrol the Switch Line and shelled the
German infantry. It was at 11 a.m. that the Whippets entered
the fray, driving from northern Cachy and into the midst of

the enemy infantry. Firing their machine guns, they scattered the Germans left and right. Captain Price had been told incorrectly that he was attacking unsupported enemy soldiers. Instead, he ran into the other two A7Vs consisting of *Schnuck*, commanded by 2nd Lieutenant Albert Mueller and *Siegfried*, under 2nd Lieutenant Friedrich-Wilhelm Bitter. The latter moved forward and opened fire along with supporting artillery.

Price's tanks had no way of defending themselves against the Sturmpanzerwagens' 57mm cannon. Suddenly one of the Whippets lurched to a halt with smoke pouring from it, and a second burst into flames. It is unclear who hit them, but this unexpected resistance had the desired effect. The rest of the Whippets then began to withdraw on Cachy, but only three made it undamaged after two of them broke down.

Lieutenant Mitchell's luck also ran out. He had already been accidentally bombed by a friendly aircraft that thought his tank was German. This error is perhaps easy to understand in light of the Germans redeploying captured Mark IVs. Spotting a German tank at a range of some 1,000 yards, Mitchell opened fire, but this drew the attention of a German mortar crew. 'We had been hit at last,' he recalled. 'We got out and made for the nearest trench some 50 yards back. It was about 12.45 p.m.' Mitchell was awarded the Military Cross for his bravery during the engagement between Cachy and Villers-Bretonneux, while his right-hand gunner, Sergeant J.R. McKenzie gained the Military Medal. Mitchell's citation read:

> For most conspicuous gallantry and devotion to duty in action against enemy tanks at Cachy on April 24, 1918. This officer was in command of a male tank in action east of the Cachy Switch Line, when hostile tanks came in action.

He fought his tank with great gallantry and manoeuvred it with much skill in order to bring the most effective fire on the enemy one, but to avoid offering a greater target than possible. As a result of his skilful handling of his tank and his control of fire, he was able to register five direct hits on the enemy tank and put it out of action. Throughout he showed the greatest coolness and initiative.

Villers-Bretonneux showed the British Mark IV and Whippet to be far more manoeuvrable than the unwieldy Sturmpanzerwagen. However, the females and the Whippets, armed with just machine guns, were unable to fight on anything like equal terms. This meant only Mitchell's male Mark IV was able to take on the three A7Vs. The world had witnessed its very first tank versus tank battle.

The Germans, who had not been enthusiastic about the tank concept, were to embrace it wholeheartedly under Adolf Hitler with the secret creation of his panzer divisions. Villers-Bretonneux sowed the seeds for Hitler's blitzkrieg. Similarly, although the aircraft had been developed into weapons of war with the emergence of the fighter and bomber, they only played a relatively minor part supporting land and naval operations. Again, under Hitler they were to be moulded into a much more potent weapon with the Luftwaffe. While the First World War had been a global conflict, the beginning of the Second World War with blitzkrieg was a harbinger of what became total war.

14

BLITZ BUT NO KRIEG:
BRITAIN 1940

On 7 September 1940 a mighty air armada of 320 Luftwaffe bombers escorted by 600 fighters attacked London. The city's defences, emergency services and infrastructure were swamped by the Nazi onslaught. London and the Thames were left ablaze. Adolf Hitler's intention was that his 'blitz' would drive the British Government, newspapers and the BBC out of London and close the banks and railways down. If any or all of these could be achieved, he hoped it would cause an involuntary collapse of the British state.

A sense of panic swept the country that day and the Invasion Alert code 'Cromwell' was issued. It took almost four hours to reach the Coastal Commands, as Britain braced itself for the inevitable Nazi invasion. The central players in this unfolding drama were Prime Minister Winston Churchill; General Alan Brooke, commander-in-chief (CinC) home forces; General Andrew Thorne, commander of 12th Corps where Hitler's main attack was expected to fall; and Air Chief Marshal Hugh Dowding, CinC RAF Fighter Command. Hitler's key henchmen were Reichsmarschall

Hermann Göring, CinC Luftwaffe; Admiral Erich Raeder, CinC Kriegsmarine (German Navy); and Field Marshal Walter von Brauchitsch, CinC German Army.

Poised to invade just across the English Channel were two entire German Army Groups totalling over forty crack divisions. Field Marshal Gerd von Rundstedt's Army Group A was to lead the assault, supported by Field Marshal Fedor von Bock's Army Group B. Rundstedt was to land between Worthing and Folkestone, while Bock would follow up by coming ashore at Lyme Bay. In all, over a quarter of a million German troops were to be landed within just four days. The first wave was to consist of 90,000 men supported by 250 submersible Panzers, 26,000 bicycles, 4,000 horses and two airborne divisions. The second wave was to bring in the all-important Panzer divisions, which were to be landed at captured ports.

The submersible panzers were Hitler's secret weapon and a special 'Experimental Staff' had been set up under General Georg Hans Reinhardt. Trials with amphibian tanks, landing craft and landing stages were carried out at Sylt, Emden, Husum and Putlos as part of the preparations. Reinhardt's experiments included submersible Panzer Mark IIIs and IVs, known as the Tauchpanzer, they were not truly amphibious but could operate at depths of up to 15m using a floating snorkel. It was intended that four battalions of these were to hit the British beaches within the first two hours of the invasion.

In July 1940 four sections of volunteer crews were trained on the island of Sylt and the Tauchpanzer were ready for trials at Putlos the following month. As well as 210 submersible medium tanks (168 Tauchpanzer III and forty-two Tauchpanzer IV) Hitler also had fifty-two amphibious Schwimmpanzer II light tanks available for the invasion. These

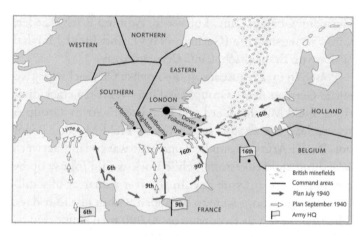

Operation Sea Lion, the proposed invasion of England, 1940.

submersibles were allocated to the 1st (Schwimmpanzer), 3rd (Tauchpanzer) and 18th (Tauchpanzer) Panzer Divisions.

Hitler's initial plans called for assaults between Ramsgate and Bexhill, Brighton and the Isle of Wight, and an isolated one near Lyme Regis. The first objective was the seizure of a line from the Thames Estuary heights, south of London, to Portsmouth. The second objective was mopping up of resistance in southern England; then the occupation of London, followed by an advance to a line from Maldon to the Severn Estuary. Beyond this, a general thrust north-wards was envisaged.

Lieutenant General Alan Francis Brooke (later Viscount Alanbrooke), CinC home forces, had the unenviable task of repelling Hitler. Just after Dunkirk, his predecessor General Sir Edmund Ironside had managed to scrape together about twenty-seven weak divisions. Only four of these were fully equipped, eight partially, while the rest were decidedly defi-cient. Ironside concentrated his troops back around Redhill,

Sevenoaks and Maidstone up to the Thames Estuary. There was an initial expectation that landings would be attempted in East Anglia. Brooke, who in France had commanded the British Expeditionary Force's (BEF) 2nd Corps and after Dunkirk the British withdrawal from the rest of the country, once in charge in England reinforced the forward areas in July.

It was under Ironside that GHQ Home Forces, based at Twickenham, on 5 June 1940 came up with the Invasion Alert code word 'Cromwell', superseding the earlier 'Caesar'. The irony that Oliver Cromwell had ruled as lord protector without king or Parliament – effectively a dictator – seemed lost on wartime planners. What they wanted was a military leader who encapsulated strength and success, and Cromwell certainly fitted the bill. Four years later, Cromwell was also to have a tank named after him; as a pious cavalry commander, he would have either been flattered or considered it a frivolous honour.

The invasion was expected in Kent and Sussex, which was the responsibility of General Thorne's 12th Corps, also tasked with defending parts of Hampshire and Surrey and the entire coastal strip from Greenwich to Hayling Island. In June, Thorne's corps consisted of a single division and a motorised brigade.

Ironside and Thorne were in an impossible situation, with the former remarking, 'I am trying to piece together an Army in the most terrible crisis that has ever faced the British Empire'. In terms of saving manpower, the evacuation of the BEF from Dunkirk was a miracle; but in terms of equipment losses it was a disaster of the first magnitude. Hitler captured almost every tank the British Army possessed and most of its motor transport. His haul included 600 tanks, 75,000 motor vehicles, 1,200 field and heavy guns,

1,350 anti-aircraft and anti-tank guns, 6,400 anti-tank rifles, 11,000 machine guns, 90,000 rifles and 7,000 tons of ammunition. By the summer of 1940, Britain's armoured units possessed just 240 medium tanks, 108 cruiser tanks and 514 largely useless light tanks.

Many British officers were dismayed at having to abandon their equipment. 'It was impossible to evacuate our heavy weapons and transport,' recalled Lieutenant Colonel Horrocks:

> so as soon as we got inside the bridgehead we were ordered to immobilise our vehicles and move on foot. The drivers hated doing this because in war each driver develops a feeling of affection for his own lorry or truck. It was a horrible sight – thousands of abandoned vehicles, carriers, guns and pieces of military equipment of all sorts. It was a graveyard of gear.

The 'Halt Order' at Dunkirk marred Hitler's successful blitzkrieg against France in May 1940 and saved the BEF from annihilation. Operation Dynamo helped 224,685 British soldiers and another 141,445 Allied troops escape the net. There were, however, sound strategic reasons for Hitler's actions. He was worried about protecting his southern flank and seizing Paris; after all, his main purpose was the defeat of France. With her armies scattered, France signed an armistice on 22 June and Hitler told General Ewald von Kleist, 'They [Britain] will not come back in this war.'

Secretary of State for War Anthony Eden visited General Thorne on 29 June, and was shocked by his complete lack of anti-tank guns. Once informed, Churchill summoned Thorne to his official country residence at Chequers in Buckinghamshire the following day for lunch. There the general pointed out that 3rd Division was his only fully trained unit and that it was to be deployed in Northern Ireland. Thorne also told the prime minister that he anticipated 80,000 stormtroopers would land on the beaches between Thanet and Pevensey. Churchill felt the Royal Navy might have other ideas, and said he had every confidence in Thorne's troops – but what else could he say under the lamentable circumstances? The British Prime Minister ordered the 3rd Division to stay put and the next day instructed General Hastings Ismay (his representative on the Chiefs of Staff Committee) to evaluate 'drenching' the invasion beaches with mustard gas. Churchill told Thorne, 'I have no scruples, except not to do anything dishonourable.'

British Auxiliary Units (designed as a stay-behind forces), some 5,000 strong, were also established under GHQ Home Forces to harass the Germans in the event of a successful invasion; they were mostly drawn from the ranks of the Home Guard. These auxiliaries were to spy on German units, kill Germans if the opportunity arose and, more controversially, kill collaborators. Life expectancy for these volunteers was short. Peter Fleming (brother of the creator of 007 James Bond) set up the first regional training centre known by the cover name '12th Corps Observation Unit'. General Thorne, in whose area Fleming's force operated, was an old friend.

Had Hitler got sufficient troops ashore, it seems Thorne would have been overwhelmed. In July, Göring had envisaged liquidating Hugh Dowding's RAF Fighter Command in four days, followed by the destruction of Brooke's ground forces in four weeks. By August, Thorne had up to six divisions available, in particular 45th Division deployed in the Newhaven-Rye area, which would have to fend off seven German divisions.

With Europe conquered, Hitler's remaining problem was to persuade the recalcitrant Churchill to submit to his domination of the continent. Hitler said of the situation, 'The British have lost the war, but they don't know it; one must give them time, and they will come round.' Admiral Raeder had two conferences with the Führer on 21 May and 20 June. During both, Hitler gave the subject little attention, obviously regarding invasion as unnecessary to force Churchill to sue for terms.

Hitler did not want the British Empire destroyed, commenting, 'This would not be of any advantage to Germany, German blood would be shed to accomplish something which would only benefit Japan, the United States and others.' Also, after the Fall of France, Hitler intended to demobilise part of the German Army, and this too seemed to indicate no invasion. However, the Führer gave Churchill one month to come to his senses.

It was apparent that Hitler's relations with Joseph Stalin, ruler of the Soviet Union, were in rapid decline and, with his designs turning east, Hitler desperately needed peace with Britain, but the pugnacious Churchill would not negotiate. Considering Churchill's completely hopeless position, this must have infuriated Hitler, who could not understand why the British Prime Minister was being so unreasonable. Faced with Churchill's continuing obstinacy, Hitler drew the problem of invading Britain to the attention of his three service chiefs, Brauchitsch, Göring and Raeder on 2 July 1940.

Fourteen days later, Hitler issued War Directive No. 16, 'On Preparations for a Landing Operation against England'. Operation *Seelöwe* (Sea Lion) officially came into being and Hitler stated threateningly, 'I have decided to prepare a landing operation against England and, if necessary, to carry it out', adding, 'The aim of this operation will be to eliminate the English homeland as a base for the prosecution of the war against Germany, and if necessary to occupy it completely.'

On 19 July, the Führer made one last appeal to Churchill via the Reichstag and promised 'unending suffering and misery' if he would not come to terms. It seems, however, that this was only blustering, because the Italian Foreign Minister, Count Ciano, noted, 'I believe that his desire for peace is sincere'. But Hitler's bluff was called when Churchill refused to play ball once more.

When a further directive was issued it did not elaborate on the invasion of Britain. No. 17 dated 1 August 1940, was 'For the Conduct of the air and sea Warfare against England'. Hitler stated, 'In order to establish the necessary conditions for the final conquest of England, I intend to intensify air and sea warfare against the English homeland.' This, in effect, meant that Göring's Luftwaffe would have to bear the brunt of the offensive against Britain.

The backbone of Göring's bomber fleet were the twin-engine Heinkel 111 medium bomber and the single-engine Junkers 87 Stuka dive-bomber. These were protected by the Messerschmitt Bf 109 and Bf 110 fighters. To fend them off, Fighter Command was equipped with the recently introduced very agile Hurricane and Spitfire fighter aircraft. The drawback with the Heinkel 111 was that its role was tactical not strategic, as it was not designed for attacking targets at long distances. Likewise, the Bf 109 was not intended for long-range escort duties. The Bf 110, although long range, was outclassed by its British counterparts.

Hitler's biggest enemy was now time. He wanted to attack Stalin, so he had insufficient opportunity to organise and conduct a full-scale invasion of Britain. Also, the longer Churchill was left, the stronger his defences and resolve would grow. Britain was weakest during May and June 1940 and Hitler's biggest mistake was to allow the evacuation at Dunkirk and then not follow up with an immediate invasion. After all, the BEF returned home almost weaponless, however a German invasion had not been planned, nor was it practical at that stage.

Göring's first large-scale air attack, *Adler Tag* (Eagle Day), took place on 13 August 1940, but went disastrously wrong. Ridiculously, no radio link existed between the fighters and bombers; as a result, the bombers attacked twenty-four RAF sector stations defending London without fighter escort. The Luftwaffe's inexperience in promoting large-scale co-ordinated aerial operations was to cost Göring dearly. Furthermore, he had significantly underestimated the British Radio Direction Finding (Radar) equipment. From the first blip on the screen to the pilots racing for their planes had a time-lapse of just six minutes, although it did take some time for Fighter Command to perfect interception to a fine art. Another advantage they had was the Ultra Intelligence intercepts, which were eavesdropping on Hitler's top-secret encrypted communications.

It is worth noting that a German High Command directive of 16 August stated, 'On D-1 day the Luftwaffe is to make a strong attack on London, which should cause the population to flee from the city and block the roads'. It was clear from the start that the air offensive on London was designed to provide direct support for the actual invasion. Thus, Hitler's decision to start bombing London on D-15 instead of D-1 was a political decision that severely hampered the Luftwaffe's contribution to the prosecution of Sea Lion.

On 24 August, London was accidentally bombed, and the following day a furious Churchill retaliated against Berlin. Eighty-one British bombers were sent, although only twenty-five reached their target. This act was to have far-reaching consequences as Hitler turned his attentions away from the RAF's airfields to Britain's cities. Also, significantly, on the afternoon of 2 September, General Kurt Student, CinC German airborne forces, was told by Göring at Carinhall, 'The Führer doesn't want to invade Britain.' To Student's response, 'Why not?', Göring shrugged, 'I don't know. There'll be nothing doing this year at any rate.' From then on, Student thought Göring was trying to force Hitler's hand.

By 6 September Göring was well, although tardily, on his way to making the invasion possible by crippling Dowding's very hard-pressed Fighter Command. But ironically, the following day Hitler foolishly diverted the whole of Göring's Luftflotten (Air Fleet) 2 and 3 from their attacks on the RAF sector stations to bomb London. He wanted retribution for the raids on Berlin and in the process changed the entire nature of the Battle of Britain. Even the German air fleet commanders, Generalfeldmarschall Hugo Sperrle and Generalfeldmarschall Albert Kesselring, were divided over how to defeat the RAF.

Dowding had built up Fighter Command since 1936 and had ensured it had the latest equipment available, which proved vital in the coming battle. Dowding also managed to save it from total obliteration during the Fall of France. Fighter Command, although used as an excuse by the Germans, fought a battle that eventually had little or nothing to do with the threatened invasion.

Sperrle sensibly wanted to continue the attacks on the RAF's fighter stations, but Kesselring reasonably argued that this was a waste of time, because the RAF could simply withdraw out of range. He observed, 'We have no chance of destroying the English fighters on the ground. We must force their last reserves of Spitfires and Hurricanes into combat in the air.' This could be best achieved by attacking London, or

so he believed. Such an early change in Hitler's strategy saved
Fighter Command from total destruction. It also meant that
Hitler and Göring had abandoned the policy that would have
best directly supported the launch of Sea Lion.

As a prerequisite for the operational success of Sea Lion,
Raeder argued complete air supremacy had to be achieved.
This, in reality, was not possible. The RAF could have been
severely damaged in the short run, as indeed it was, but it
would have taken a concentrated and protracted campaign to
smash it completely. Even if the attacks on RAF facilities had
been maintained, it seems likely that Dowding would simply
have withdrawn to the safety of the Midlands, out of the
range of the Luftwaffe based in northern France. He would
have still been able to operate over southern England and the
edge of the Channel, thus affecting any invasion.

Initially the air offensive was in direct support of Sea Lion
with the destruction of the RAF, but then Sea Lion became
contingent on the Luftwaffe's success in beating London into
submission. General Alfred Jodl stated as early as 30 June:

> A landing in England can only be contemplated when air
> supremacy has been established. A landing should there-
> fore not be undertaken in order to accomplish the military
> defeat of England, which can be practically achieved by
> the Luftwaffe and the Navy, but only for the purpose of
> administering the Coup de grâce …

Thus, responsibility for the success of Sea Lion fell on Göring
and Raeder, a task that neither really had the ability to achieve.

In the meantime, desperate to stave off invasion, Churchill
stepped up his attacks on Hitler's gathering invasion fleet. Air
Marshal Arthur 'Bomber' Harris, who was deputy chief of
air staff at the time, recalled:

What Hitler wanted was protection from air attacks for a seaborne invasion and at the time our fighters could not have been a serious threat to shipping. In those days there were no rockets or bombs on our fighters, and the protection he wanted was therefore against bombers.

This is not quite true. Fighter Command's planes would have been very effective against the unarmoured invasion barges, or at least the passengers and crew. Harris disturbed rather than disrupted Hitler's preparations. Bomber Command's largely inaccurate bombers were not very effective, although Harris claimed:

It was definitely Bomber Command's wholesale destruction of the invasion barges in the Channel ports that convinced the Germans of the futility of attempting to cross the Channel ... our attacks on the barges began in July 1940, well before the main air battle of Britain developed, and were highly successful.

During September, Hitler's menacing invasion fleet moored in French and Belgian harbours accounted for about three-quarters of Bomber Command's total effort, with over 1,000 tons of bombs dropped. Out of 3,494 craft assembled for the invasion, the RAF only actually damaged or sank 243. Nonetheless, important docks and supplies were also damaged, while Raeder's schedules were severely hampered.

Churchill's naval strength was also considerable, to counter the first assault he had some seventy-nine destroyers (there were nine in the Humber, nine at Portland, eight at Rosyth, nine at Dover, twelve at Portsmouth and thirty-two at Plymouth). How many of these ships the Luftwaffe could have dealt with is debatable, though without the RAF they would have been very vulnerable to air attack. In addition, not all these vessels would have been available at the same time. German deception plans for an invasion of the east coast, Operation *Herbstreise* (Autumn Journey), did help keep the Home Fleet at Scapa Flow.

In the north at the beginning of September there were seven capital ships, twelve cruisers, seventeen destroyers and an aircraft carrier at Scapa – quite a daunting force. There were also two capital ships at Portland. Although the Royal Navy made aggressive patrolling in the Channel, they were reluctant to risk attacking shipping in French ports, for fear of damaging their strength; this job was left to Harris.

From the very start, Admiral Raeder was against the invasion due to his lack of naval strength. On 10 August 1940, a naval memorandum noted, 'Even in 1941 the Navy's strength will not be sufficient to exercise a decisive influence on the ratio between our forces and the enemy's.' German naval weakness had been severely aggravated by Raeder's losses during the invasion of Norway. The Germans had lost thirteen warships sunk and three battleships damaged. Throughout July, Raeder opposed Sea Lion, feeling the lack of naval supremacy would hamper Hitler's Channel operations and render their ability to provide mine-free channels for the invasion fleet dubious. The Royal Navy would simply attack German minesweepers and re-mine. Coupled to this, Raeder doubted his ability to defend the invasion corridors.

Furthermore, there was the problem of assembling sufficient transports as the process itself was disrupting the German economy. Raeder's staff calculated he needed 1,722 barges, 471 tugs, 1,161 motorboats and 155 transports – some 3,509 vessels. For D-Day in June 1944, the Allies used almost 7,000 craft. Also, due to the urgency of the situation there was insufficient time to construct proper assault landing craft, transport ships or specialised armoured fighting vehicles, all indispensable features of the Allied landings in Normandy and the Pacific. Instead, Hitler had to rely on slow European barges with no assault capabilities whatsoever. Proper landing craft were designed, but were not developed in time for the invasion.

Hitler's U-boat fleet numbered only fifty-seven (not all of which were suitable for operations in the Atlantic) and Admiral Doenitz, U-boat CinC, stated he would require about 300 for an effective blockade of Britain. They were not produced in sufficient numbers until the autumn of 1941 when the Battle of the Atlantic really got under way. Hitler had estimated that in July an effective blockade of Britain would take one or two years. He was right – it did not reach serious proportions until 1942.

Raeder saw a wide-front invasion as suicide, while von Brauchitsch regarded a narrow-front invasion equally as suicidal. Eventually Hitler compromised, reducing the front to 90 miles. The first plan, however, was a true invasion operation and its rejection can be seen as significant. Of the reduced front, von Brauchitsch said, 'It was a frontal attack against a defensive line, on too narrow a front, with no prospects of surprise and with insufficient forces reinforced in driblets.'

On 31 July 1940, Sea Lion was scheduled for 15 September, but by July Hitler's plans were well under way for the attack on Stalin. He remarked, 'Stalin is flirting with Britain to keep her in the war and tie us down, with a view to gaining time and take what he wants, knowing he could not get it once peace breaks out.'

Hitler did not let the invasion plans for the Soviet Union interfere with his war against Churchill. On 4 September Hitler publicly pledged himself to the invasion, although it was postponed from 15 September to 21 September on the request of Raeder. Hitler had undertaken to ratify the date of D-Day ten days in advance to allow for mine clearing and laying operations.

In Britain, everything pointed to an invasion being imminent. On 1 September, RAF photographic reconnaissance discovered hundreds of invasion barges heading towards the Scheldt Estuary and Dover Straits. The numbers photographed gathering between Flushing and Le Havre grew. There had been just eighteen barges in Ostend on 31 August; by 6 September alarmingly there were over 200. The concentration of shipping along the French and Dutch coasts was so great that General Brooke issued the preliminary Invasion Alert – 'Attack probable within the next three days'. He also visited the 1st Armoured Division to assess their preparations, followed by a meeting with Churchill.

On 7 September, Brooke recalled in his diary, 'All reports look like invasion getting nearer. Ships collecting, dive-bombers being concentrated, parachutists captured … sending out order for "Cromwell" State of Readiness in Eastern and Southern Commands.' This signalled Alert No. 1, 'Invasion imminent and probable within twelve hours'. The huge Luftwaffe attack on London that day seemed to confirm everyone's greatest fears.

In fact, Brigadier John Swayne, deputy chief of staff at GHQ Home Forces, unable to locate his boss, issued the alert at 8.07 p.m. to the Eastern and Southern Commands: 4th and 7th Corps (in GHQ Reserve) and London District. It was then repeated to all other commands 'Cromwell' immediately caused widespread confusion. The 7th was a Saturday, which meant most headquarters were largely empty and, to many duty officers, 'Cromwell' either meant nothing at all or that the invasion was actually under way.

Panic rapidly set in. Across the villages stretching from Portsmouth to Swansea the Home Guard began to ring the church bells, roadblocks were closed, as were the telephone lines; they then stood to all night (as they always did) in anticipation of Mr Hitler's arrival. In the Southampton area, men of the 4th Division dozed in coaches, rifles at the ready. On the roads leading to Canterbury, Maidstone and Horsham, soldiers manned massive 600-gallon flamethrowers ready to incinerate the German blitzkrieg at 500 degrees Fahrenheit. In Eastern Commands' area, overzealous Royal Engineers even blew up several bridges.

At Middle Wallop, men of RAF 609 Squadron, like many others, sat strapped in their cockpits, engines turning, facing downwind ready to go. Pilot Officer 'Nobby' Clarke, serving with the Gosport Army Co-operation Station, was ordered to 'get cracking – light all points …'. Flying his Skua target-towing plane, he was to dive-bomb the petrol pipelines with incendiaries to set ablaze the inshore waters and invasion beaches. Clarke was heading east towards Littlehampton when he was recalled.

The following day Brooke observed, 'All reports still point to the probability of an invasion starting between the 8th and 10th of this month. The responsibility … is a colossal one … There is nothing to be done but to trust God and pray for his help and guidance.' For the next week Hitler's bombers relentlessly attacked London day and night.

'No one,' Churchill broadcast on the 11th, 'should blind himself to the fact that a heavy full-scale invasion of the island is being prepared with all the usual German thoroughness and method, and it may be launched now …' Hitler should have issued an order confirming or postponing Sea Lion that day, but he did not. His failure to take a decision then was the first instance of his wavering in the execution of Directive No. 16.

Churchill, Brooke and Thorne waited for the invasion that mercifully did not come.

Three days later, the German Naval War Diary recorded, 'According to the state of preparations today the execution of the operation by 21.9.40 as previously arranged, is thus ensured.' Significantly, it further added, 'The Führer thinks the major attacks on London may be decisive, and a systematic and prolonged bombardment of London may result in the enemy's adoption of an attitude which will render Sea Lion superfluous.' It was wishful thinking.

Even by the end of August, German General Putzier said an airborne landing was out of the question. The operational side of the airborne invasion, involving a division landing on the South Downs and north of Dover, was rapidly scrapped due the erection of increasing numbers of British anti-airborne obstacles. Student was in hospital at the time, but his plans for a successful operation would have been more ambitious, seizing airfields further inland supported by airlifted infantry divisions. This sounded somewhat like the hamstrung strategy used in Norway. Student returned to duty in September, but showed no personal interest in Sea Lion.

Throughout the summer, as Hitler continued to stockpile airborne equipment in northern France, the fear of German paratroops and 'Nuns in hob-nailed boots' prevailed in Britain. The 1942 movie, *Went the Day Well?* aptly highlighted this fear of fifth columnists and invasion, with a story about German troops disguised as Royal Engineers arriving in a sleepy English village. The plucky locals hold out until reinforcements arrive. Ultimately, the success of any German airborne operation is questionable in the subsequent light of the high casualties incurred during Operation Mercury, the airborne invasion of Crete in 1941. In one day, about half the

10,000 German paratroops dropped were killed and some 250 Junkers transport planes lost.

In Britain, everyone remained anxious. On the 15th Brooke recalled the terrible tension, 'Still no move on the part of the Germans. Everything remains keyed up for an early invasion, and the air war goes on unabated. This coming week must remain a critical one ... The suspense of waiting is very trying ....' Although morale was good, Brooke wished he had another six months to finish equipping and training his twenty-two divisions, who were holding a front twice that which had been defended in France by eighty British and French divisions.

On 14 September 1940, Hitler deferred his decision for three days, pushing the invasion back to 27 September, which was outside the invasion period stipulated by Raeder. Nor were the weather forecasts for the end of the month encouraging. Göring had not attained air superiority, despite the continuing air offensive. Dowding's Fighter Command had not been defeated, Harris' Bomber Command and the Royal Navy were still operating and under Brooke the British Army's strength was growing daily. The British publics' unwavering resilience to the Blitz also confounded Hitler and his senior commanders.

From August through to September Göring had lost 1,140 aircraft. Another large-scale attack was conducted on London on 15 September. This was also to fail in breaking the back of the RAF, and Luftwaffe losses continued to mount alarmingly. The following day, Göring called a conference to inform the Luftwaffe that the escort fighters had failed to smash their opponents. It was announced that 'the enemy air force is still by no means defeated: on the contrary it shows increasing activity. The weather situation as a whole does not permit us to expect a period of calm.

The Führer has therefore decided to postpone Operation Sea Lion indefinitely. This heralded the end for Sea Lion.

Hitler must have been well aware that an invasion on 27 September was not in the slightest way a serious proposition. By 17 September, the German High Command knew that Hitler had made up his mind. Preparations continued, although by the 19th the RAF had finally managed to force Hitler to disperse his invasion fleet. At that point 'Cromwell' was rescinded, much to the relief of Brooke and his anxious commanders who were thoroughly fed up with its disruption. Nonetheless, it was not until 12 October 1940 that Hitler officially called off the invasion and it was not until 13 February 1942 that the troops earmarked for Sea Lion were finally released. Thankfully, code word 'Cromwell' proved a false alarm.

15

Storming the Beaches: D-Day 1944

Sometime after 1 a.m. on 6 June 1944 a very bleary-eyed Admiral Hoffman was roused from his bed at the Headquarters of Chief of Operations German Naval Group West in the Bois de Boulogne, Paris. Chief of Staff Hoffman found himself leafing through a series of reports from their remaining naval radar stations. Despite the Allies' best efforts, there could be no hiding the vast fleet approaching the Normandy coast and Hoffman turned to his men, 'This can only be the invasion fleet. Signal to the Führer's headquarters the invasion is on.'

At 2 a.m. the first British and American parachutists, the vanguard of a vast aerial armada supporting Operation Overlord, drifted from the Normandy night skies. In addition, gliders were released from their tow-planes into the darkness as pilots grappled with the controls, trying to ensure a safe landing. D-Day and the Allied liberation of Nazi-occupied northern France had commenced.

On the western flank of the Allied assault on Hitler's *Festung Europa*, about 920 transport aircraft and 100 gliders of the United States Army Air Force (USAAF) dropped the US 82nd and 101st Airborne Divisions in the Sainte-Mère-Église area of the Cherbourg Peninsula, just inland from the Utah invasion beach. Despite being scattered, they still managed a good job. On the far eastern flank, about 360 aircraft and ninety-six gliders delivered the British 6th Airborne Division between the River Dives and Orne, inland from what was designated Sword beach.

The very first Allied landing in Normandy was achieved by six British Horsa gliders, towed by Halifaxes and bearing six platoons of the 2nd Battalion of the Oxford and Buckinghamshire Light Infantry plus a unit of Royal Engineers. Their task was to grab the two bridges over the Caen Canal and the River Orne. Four of the gliders came down with great accuracy, PF800 touched down within just 47 yards of the Caen Canal swing bridge. Unfortunately, a fifth landed half a mile away and the sixth on a bridge over the River Dives, some 7 miles away.

To many, these air landings heralded the start of Overlord but, in reality, the concerted air attack had started way before this. In fact, the Allies had gained air superiority over Hitler's Luftwaffe in 1943 and never relinquished it. From that point, under Operation Pointblank, the RAF and USAAF had conducted preliminary operations that would contribute to Overlord's success. Key amongst these was concentrating heavy bomber attacks on Germany's fighter plants. This forced the Luftwaffe to draw forces away from the front line to defend the factories. What followed was a bloody battle of attrition that the German pilots were slowly losing, although fighter production ironically continued to increase.

Then, at the beginning of 1944 Pointblank was redirected to support the Overlord Air Plan. The 'Heavies' began bombing France's railways and marshalling yards, while the medium and light bombers attacked the strategic bridges over the Seine and Loire, as well as various coastal radar sites. The intention was to prevent the Germans moving reinforcements into France and to blind them along the English Channel. Many of these attacks went largely uncontested, with anti-aircraft fire posing the greatest danger to Allied pilots.

Luftflotte 3, stationed in France, only had 300 aircraft, of which fewer than 100 were fighters. In the event of invasion this force was supposed to be reinforced by 600 aircraft redeployed from air defence duties in Germany. The situation for the Luftwaffe in northern France was grim, with its airfields being bombed and its radar blinded, it had lost many of its experienced crew either over Germany or Russia and the training programme was not turning out adequate replacements. RAF Air Vice Marshal H. Saunders informed his No. 11 Group, 'The enemy is not expected to react in the air in great numbers on the first day, but may appear in force up to two or three hundred strong in three or four days. That will be your opportunity.'

The reality was that by 1944 Hitler's Luftwaffe in the west did not even have the right aircraft with which to repel an Allied invasion backed by overwhelming air superiority. The Luftwaffe in France really needed several squadrons of fast reconnaissance aircraft equipped with photographic equipment that could elude enemy fighters, numerous squadrons of rocket-firing ground-attack aircraft that could pound the invasion beaches, plus similar numbers of fighter-bombers and tank-busting aircraft that could set about the Allies' tanks. Even if the Luftwaffe had the luxury of such equipment, it would have struggled to deploy it as its facilities were systematically targeted.

While the Germans lacked adequate air cover, the one thing they had in abundance was anti-aircraft guns or flak guns. By 1944 the Luftwaffe had a total of twenty-six flak divisions. Some of these were organised into three corps, 1st and 2nd Flak Corps were deployed on the southern sector of the Eastern Front, while the third was in France. Two new corps came into being in the second half of the year, but these were used to defend Germany from the Allies' heavy bombers.

The 88mm dual-role Flak 36 anti-aircraft gun posed a real threat to Allied aircraft, and most of those deployed in Normandy came under General Wolfgang Pickert's 3rd Flak Corps. This was responsible for providing air defence cover for Army Group B, charged with defending northern France and the Low Countries, and Army Group G, holding southern France. In addition, most German divisions had their own dedicated anti-aircraft units.

Once the Allies were ashore, standing orders that flak artillery should not be used to support the ground troops were understandably ignored. When Panzergruppe West was activated, 3rd Flak Corps was placed under its command, although not all of Pickert's batteries were released for frontline duty. As the fighting progressed, the priority for the German flak guns was defending the Caen–Falaise road and then the line of retreat to Rouen from air attack.

By mid-1944, the Allied strategic air forces, consisting of RAF Bomber Command, RAF Coastal Command and the USAAF 8th and 15th Air Forces, could field over 5,000 four-engined heavy bombers against Germany and the occupied territories. In addition to this mighty heavy bomber fleet, the Allied tactical air forces numbered 2,840 fighters and fighter-bombers, plus 1,520 light and medium bombers. The Allied Expeditionary Air Force (AEAF) comprised of the RAF 2nd Tactical Air Force and the US 9th Air Force. Also subordinate to the invasion were the forces of RAF Air Defence of Great Britain (ADGB).

Out over the shipping lanes in the English Channel, six groups of P-38 Lightnings from the US 8th and 9th Air Forces were given responsibility for protecting them by day, while at night Beaufighters and Mosquitos of RAF ADGB and No. 85 Group took responsibility for the night-fighter role. The anchorages and beaches from 5 miles inland to 15 miles off shore were protected by a low cover of Spitfires and a higher cover of P-47 Thunderbolts.

Hitler's defensive strength on all fronts prior to the Normandy landings consisted of 1,523 single-engined fighters of which 59 per cent were serviceable; 778 night-fighters of which 68 per cent were serviceable; 242 twin-engined fighters of which 51 per cent were serviceable and 1,005 ground-attack aircraft of which 75 per cent were serviceable. These aircraft, however, were dispersed across the vast Eastern Front, Germany, Italy, Belgium, the Netherlands and France. To stretch the Luftwaffe's resources even further, they were acutely conscious that an Allied landing in the south of France along the Riviera was also a distinct possibility. Dubbed Operation Dragoon, an Allied assault was to take place there in August 1944.

Four days after D-Day commenced, Luftwaffe strength was listed as 1,300 aircraft (excluding transport planes) deployed in France and the Low Countries, of which just over half were available for operations. This force consisted of 475 single-engined fighters with 290 serviceable; 464 bombers with 267 serviceable; 170 night-fighters with ninety-six serviceable and forty-six twin-engined fighters with twenty-six serviceable. In light of the vast Allied air fleets deployed against them it was too little too late.

The airlift required to drop three whole airborne divisions into Normandy was enormous and was provided by USAAF's 9th Troop Carrier Command and the RAF's Nos 38 and 46 Groups, which came under the US 9th Air Force and RAF 2nd Tactical Air Force respectively. By May 1944, 9th Troop Carrier Command had 1,167 aircraft at its disposal, plus over 100 CG-4A Waco gliders, that could each carry fifteen troops. Transport aircraft were the C-47 Douglas Skytrain and C-53 Skytrooper – these were essentially the same, but the former could be converted into a freighter; both could be used for glider towing. By 1 June 1944, the RAF's Nos 38 and 46 Groups had 406 aircraft ready for operations along with 1,120 gliders, of which seventy were the big Hamilcars capable of carrying forty troops or up to 8 tons of freight; the rest were Horsas, which could carry twenty-nine troops or 3 tons of equipment.

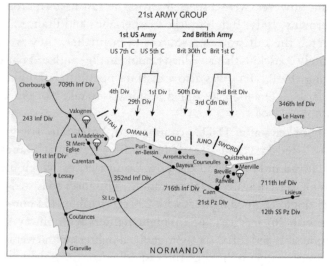

Operation Overlord, D-Day, 6 June 1944.

These aircraft and gliders had to shift the US 82nd and 101st and the British 6th Airborne Divisions. The US airlift was given priority because they had the task of seizing the crucial exits from the Cotentin Peninsula, south of Cherbourg. To deceive the remaining enemy radars, RAF Stirling bombers dropped metal foil giving a false radar image known as 'Window' in an area south of the actual location. The C-47/53 fleet lifted 17,000 men and flew from west to east across the Cotentin Peninsula. The 101st were to capture the landward exits from Utah beach, to destroy the bridges on the road leading to Carentan and to protect the southern flank of the invasion forces

Wing Commander Roland Beaumont, at RAF Tangmere, recalled how just before D-Day they were briefed by the commander of No. 11 Group:

The animated buzz of the conversation hushed suddenly as the Air Officer Commanding, Air Vice-Marshal Saunders, entered the Operations room. He waited while a sheaf of maps was unrolled and pinned to the blackboards. This was IT, right enough. Red, blue, yellow and black lines stretched across the maps from London, Dungeness, Portsmouth, Southampton and the West Country, to a point south of the Isle of Wight, and then across the Channel to Normandy.

There was no need to command attention. Every man waited upon the AOC in eager anticipation.

Gentlemen, he said, tomorrow is D-Day. I will outline briefly the plan as it concerns us, and then I wish you to return to your units to open and study your sealed orders for Operation Overlord. You will brief your wings at dusk and be instantly available from midnight onwards.

Before the American, British and Canadian assault forces came ashore, German defences were drenched in 1,760 tons of bombs. These attacks were conducted by B-17 Flying Fortresses and B-24 Liberators of the USAAF's 8th Air Force, and B-26 Marauders of the 9th. Unfortunately, the concrete protecting the German coastal guns proved largely impervious to the air and sea bombardment. As the warships approached to conduct their sea-to-shore bombardment, they were shrouded from coastal retaliation by a smoke screen laid by RAF Boston squadrons of No. 2 Group.

Implementing Overlord required one of the largest seaborne invasion fleets in history, comprising 1,213 warships, 4,126 landing ships and landing craft, and 1,600 other vessels – almost 7,000 vessels in total. The naval element of D-Day came under the designation of Operation Neptune. The key craft for getting the tanks ashore were the Landing Ship Tank (LST), which was capable of carrying sixty tanks/vehicles or 300 troops. In total, 236 were committed to Neptune along with 768 landing craft tanks (LCTs) and forty-eight landing craft tanks (armoured). Specialised landing vehicles included 414 amphibious Sherman Tanks (duplex drive).

Once all these craft had been assembled, the really big headache for the planners was the loading and landing schedule. General Francis de Guingand, chief of staff of Montgomery's 21st Army Group, recalled:

> Many of the ships carrying vehicles and equipment, although not required at once, had to be loaded before D-Day. Relative priorities so much depended upon the course of the campaign, and therefore decisions at this early date were not easy. For instance, by what date would another load of bridging be required? This depended, of course, upon our rate of progress. When would the

12th US Army Group take over? If we shipped over all
their vehicles too soon, valuable space might be wasted.
Such problems came up each day.

Another major concern was the appearance of numerous
deadly beach obstacles in the selected landing zones, which
posed a threat to the smaller assault craft. These consisted of
steel and wooden posts, many of which had mines attached,
capable of tearing a craft's hull open. The navy conducted
various experiments to determine their effect on the different types of landing craft. General Hobart's 79th Armoured
Division, which operated the specialised armoured assault
vehicles known as the 'Funnies', was given the task of clearing the way.

Due to the different tide times and bombardment lengths,
the invasion beaches, stretching from La Madeleine in the
west to Ouistreham in the east, had their landings staggered.
The American Utah and Omaha were assaulted at 6.30 a.m.,
while the British Gold and Sword beaches were at 7.25 a.m.,
and lastly the Canadian Juno at 7.45 a.m..

Utah, forming the far western flank centred roughly
on La Madeleine, was attacked by General Collins' US 7th
Corps, led by the US 4th Infantry Division. Their job was
to link up with the 82nd and 101st Airborne, establishing a
bridgehead over the River Vire and the nearby canal ready
to link up with Omaha to the east. Due to the tide, the
American GIs went ashore 1,000 yards south of their landing zone. Twenty-nine Sherman DD tanks spearheaded the
assault and were launched 5,000 yards from the shore. Little
resistance was encountered, consisting mainly of small-
arms fire. By 8 a.m. Pouppeville was attacked, and the US
4th Infantry managed to push 4 miles inland, brushing aside
most of the German resistance. At the end of the day, they

had successfully put ashore 23,000 troops, 1,700 vehicles and 1,700 tons of stores.

The US 5th Corps, led by the US 1st Infantry Division, landed on Omaha beach bordered by Vierville-sur-Mer and Sainte-Honorine. The preliminary bombardment lasted only 40 minutes and consequently many of the German defences remained intact. Also, the shingle beach was bordered by marshland and a high bluff, making it an ideal fire zone. Because of the rough seas and enemy fire, out of thirty-two vanguard DD tanks, only five cleared the beach, while out of the fifty-one tanks landed dry-shod by the assault craft, eight were knocked out before even clearing the sea. Under heavy machine gun, mortar and artillery fire, the Americans were cut to pieces as they staggered from the water and, denied armoured support, they were unable to clear the beach.

The Americans had declined the offer of 79th Armoured's 'Funnies', and German fire was so intense that out of the engineers' sixteen bulldozers put ashore on the right side of the beach, only two were serviceable. By 9 a.m., a few Americans had reached the top of the bluff and were beginning to move inland towards the villages. The GIs suffered an appalling 2,500 casualties, and had only managed to get 2 miles inland, but by nightfall 33,000 men were ready for the next round of fighting.

The British and Canadian eastern task force attacked a broad 25-mile front, between Port-en-Bessin and Ouistreham. Gold beach, centred on Le Hamel and La Rivière, was assaulted by the British 30th Corps led by the 50th Infantry Division. Their task was to take Port-en-Bessin in order to link up with the US 5th Corps, thrust for St Leger on the Caen–Bayeux road and seize Bayeux.

At 7.25 a.m., assault units from 79th Armoured, consisting of Sherman Crabs (flail tanks) and Churchill AVREs

(Armoured Vehicle Royal Engineers), went in. Once again, due to the rough sea, the DD tanks had to be landed dry-shod; also, to increase problems the tide rose thirty minutes early. The AVREs were late and Le Hamel proved to be heavily defended. The local sanatorium had been converted into a German strongpoint and German artillery fire swept the beach. However, by the afternoon Port-en-Bessin had been taken.

By 9 p.m. Arromanches had fallen, but the drive on Bayeux had stalled, even though it had been largely abandoned by the Germans. Also, the route west from Caen had been captured, but at the end of the day a 6-mile gap existed between Gold and Omaha. About 25,000 men were put ashore and 50th Division had punched 6 miles inland.

At Juno, the assaulting formation was the Canadian 3rd Infantry Division under the British 1st Corps. The beach was centred on Courseulles and Bernières. The Canadians were to seize the two towns and drive for Carpiquet Airfield, west of Caen. In order to ensure the sea carried the landing craft over the reefs, the assault was timed for 7.45 a.m., but because of the rough sea it went in at about 8 a.m. While they did get over the reefs and most of the beach obstacles, the return trips were hazardously disastrous.

Only twenty-nine DD tanks were launched, with twenty-one reaching the shore, the rest had to be landed dry-shod from the landing craft. B Squadron, 1st Hussars Regiment, supporting the Reginas, landed at Courseulles at 7.55 a.m. with fourteen of nineteen tanks. Seven A Squadron tanks landed a few minutes after the Winnipeg Rifles, on the beach west of Courseulles. Five more tanks were landed by an LCT that earlier had had problems with its ramp.

Arriving before their armour, the Canadian infantry found many of the German positions intact. Under intense small-arms fire, they could not get off the beach and many Canadians were cut down trying to reach shelter behind a defence wall at the rear of the beach. Lacking armoured support, the Canadian infantry faltered, but an AVRE managed to blow a hole in the 12ft-high sea wall, and they began to move inland. By the end of the day, 21,500 men had been landed and the beach linked with the British 50th Division at La Rivière.

Sword Beach, centred on Lion-sur-Mer, was assaulted by the British 1st Corps, led by the British 3rd Infantry Division. Their main objective was to seize the city of Caen, the German's regional HQ, and link up with the bridgehead over the River Orne. H-Hour was 7.25 a.m. and the spearhead DD tanks were launched 5,000 yards from the shore – out of thirty-four successfully launched, only three were lost. By the evening, 29,000 men were ashore in the Sword area.

Although the Americans attacking Omaha paid a heavy price, on Utah things fared better, in part due to a preparatory attack by about 270 B-26 Marauders of the US 9th Air Force, which bombed their targets from heights ranging from 3,000ft to 7,000ft. Once the troops were ashore battling their way inland they were provided tactical support. For example, at Sword the assault was followed by RAF Typhoons that were tasked to attack enemy strongpoints and defended batteries at every opportunity. The Luftwaffe proved to be a no-show. Just two pilots Oberst J. Priller and Sergeant H. Wodarcsyk, flew over the invasion fleet in their FW 190s and shot up Sword Beach.

Once Overlord was under way, the AEAF continued to provide vital support. For the rest of D-Day, fighter-bombers of the US 9th Air Force and the RAF 2nd Tactical Air Force flew close air missions for the assault troops, either by responding for requests for help or by attacking targets of opportunity. Inland, P-51 Mustangs and P-38 Lightnings of the USAAF flew area patrols to screen the battle zones from the rest of France. During D-Day these fighters conducted almost 2,000 sorties.

The weather that had led to the postponement of D-Day continued to dog the operations of the heavy bomber force. Many aircraft from a fleet of 500 Liberators and Fortresses returned from their second mission during D-Day without bombing the eight key centres in Normandy due to poor visibility. However, this was somewhat compensated for during the night, when some 977 Lancasters and Halifaxes of Bomber Command hit nine centres with the aim of cutting the lines of communication through Normandy's main towns.

By the end of D-Day, the Allies had landed as many as 155,000 troops, 6,000 vehicles, including 900 tanks, 600 guns and about 4,000 tons of supplies. Quite remarkably, within five days 326,547 men, 54,186 vehicles and 104,428 tons of supplies had been brought ashore. Just four LCTs were lost during Neptune and the landing ship *Empire Broadsword* was sunk by a mine off Normandy on 2 July 1944. The LSTs brought 41,035 wounded men back across the English Channel by the end of September 1944.

To defeat an Allied invasion, CinC West Field Marshal Gerd von Rundstedt favoured the 'crust-cushion-hammer' concept, the crust being formed by the static sea defences, the cushion by infantry reserves and the hammer by the panzer divisions held further back. General Geyr von Schweppenburg, commander of Panzergruppe West, agreed with Rundstedt, believing the Panzers should be kept inland ready to encircle the Allies as they advanced on Paris.

Field Marshal Erwin Rommel, commanding German forces in Normandy, wanted his panzers well forward to deal with the Allies as soon as they waded ashore. He felt any airborne landings in the rear could be easily dealt with by those troops who were to hand. Rommel had made his name as a panzer leader in France and North Africa and had also orchestrated the successful seizure of northern Italy. After his experiences in North Africa, he knew only too well how potent Allied air power could be, which is partly why he advocated keeping the panzers near the coast. He did not reckon with the Allies' naval gunfire, which would pound the panzers even when they did get near the beachhead.

The dispute between Rundstedt and Schweppenburg, on the one hand, and Rommel on the other resulted in an unwieldy compromise, with Rommel retaining command of 2nd (beyond the Somme), 21st and 116th Panzer Divisions (beyond the Seine), and the 1st SS and 12th SS Panzer Divisions and Panzer Lehr remaining under von Rundstedt's authority. As a result, during the landings vital time was lost getting authorisation to commit the Panzers.

D-Day heralded the Battle of Normandy, which proved an unmitigated disaster for Hitler's panzers. Ironically, the one major advantage Panzergruppe West had over the Allies was the qualitative edge of its tanks. On the whole, the German armour deployed in northern France was vastly superior to that of the Allies and easily outgunned them. While the Allies sought to counter the Germans' technological lead on land, sea and air at every single stage of the war, their failure to develop a war-winning battle tank was a glaring mistake. The Germans realised they could never match the Allies' numbers but they ensured that they could out shoot them.

The most common Allied tank was the American built M4 Sherman. Mechanically reliable, it was handicapped by thin armour and a gun lacking sufficient punch so its good cross-country speed and higher rate of fire could not make up for its two key shortcomings. Tank crew survival was paramount, tanks could be replaced relatively easily but not experienced crews. However, the Sherman had a nasty habit of burning when hit and if this happened the crew only had a 50 per cent chance of survival.

Just four days after the D-Day landings, RAF Typhoons swooped down out of the skies onto German units deployed at a château and neighbouring orchard. They were followed by Mitchell light bombers of the 2nd Tactical Air Force. Surprised German radio operators and staff officers caught in the open scattered in all directions as the ground shook beneath them. When the severe bombing and strafing, lasting several hours, was finally over a German general, Panzergruppe West's Chief of Staff Ritter von Dawans, lay dead along with twelve other fellow officers. Schweppenburg, the commander of Panzergruppe West was wounded. In one fell swoop, Hitler's panzer forces in Normandy had been successfully decapitated by the AEAF.

Intent on resisting the D-Day landings, Panzergruppe West became operational on 8 June 1944 at Château La Caine, south-west of Caen; within two days it had sealed its own fate. Allied signal intercepts from four large radio trucks parked in nearby trees were its undoing, tipping off the Allies' fighter-bombers to its exact location. Rommel visited that day and narrowly missed the attack by just thirty minutes.

Overlord's momentum, however, could not be sustained as the weather began to deteriorate and on 19 June a massive storm halted all shipping in the Channel for three days. The two Mulberry artificial harbours were smashed by 21 June and the one off Omaha was written off and used to repair the British one at Arromanches. The build-up virtually ground to a halt, delaying 20,000 vehicles and 140,000 tons of stores. However, by the end of the month over 850,000 men, 148,000 vehicles and 570,000 tons of supplies had been successfully landed, taking the battle far into the Normandy countryside.

By late June, there were almost eight panzer divisions between Caen and Caumont on a 20-mile front facing the British 2nd Army. In particular, the 2nd, 12th SS, 21st Panzer, Panzer Lehr and the 716th Infantry Divisions were all tied up in the immediate Caen area. Facing the British were approximately 725 German tanks, while on the American front there were only 140. Caen became the bloody fulcrum of the whole battle; here the cream of Panzergruppe West was ground down in a series of unrelenting British attacks culminating in the armoured charge of Operation Goodwood.

The desperately needed German infantry divisions that could have freed up the panzers for a counter-stroke remained north of the Seine. Hitler held them back because he feared an attack across the Pas-de-Calais. By the end of June, it was evident that von Rundstedt's 'crust-cushion-hammer' tactics had failed despite the slowly increasing number of Panzer divisions; in the face of Allied firepower, tied down, the Panzers could do little more than fire fight as the situation developed. To make matters worse, by the beginning of July the unrelenting operational commitment of the Panzers was taking its toll. Rommel's luck ran out on 17 July 1944 when Typhoons caught his staff car on the open road and he was hospitalised. As the fighting moved further inland, the AEAF was unable to rest on its laurels, but by this stage the fighting had entered a new phase.

Once the American Army had broken out and the British and Canadian Armies were pressing on Falaise the fate of the panzers in Normandy was sealed. Second Lieutenant Stuart Hills, Nottinghamshire Sherwood Rangers Yeomanry, 8th Armoured Brigade, followed the British 11th Armoured

Division through the Falaise Pocket via Chambois to L'Aigle. There, he witnessed the devastation:

> The scenes in the Falaise Pocket, where Allied air power had wreaked such destruction, were horrendous. The various German divisions had a terrible pounding in the Normandy battle, Panzer Lehr, for instance lost all its tanks and infantry units, while about 50,000 of the enemy had been killed and some 20,000 taken prisoner. Thousands still lay unburied within the pocket: the roads and fields were littered with German dead in various stages of decomposition.

The exhausted panzer divisions lost all their tanks in northern France; in fact, from an accumulated tank force of 1,804 just eighty-six remained. Similarly, the independent tank battalions and assault brigades from an accumulated strength of 458 could scrape together forty-four vehicles. The panzer divisions' manpower totalled about 160,000 men during the campaign and while they had lost almost 62,000, crucially 98,000 men escaped to fight another day. The German Army had lost forty-three divisions by September 1944, roughly thirty-five infantry and eight Panzer, two more than were originally stationed in northern France. They suffered a total loss of 450,000 men – 240,000 casualties and 210,000 prisoners, as well as losing most of their equipment – 1,500 panzers, 3,500 pieces of artillery and 20,000 vehicles.

Much heated debate has raged about the effectiveness and employment of the Panzer divisions in Normandy, although it was the numerically superior German infantry divisions that bore the brunt of the fighting. The failure of the Panzers to launch a decisive counter-stroke has been blamed on a muddled chain of command, inertia, Hitler's intransigence and the Allies' superiority on the ground and in the air. The reality is that from the very start there were insufficient German armoured formations in Normandy and, although they rose to almost a dozen, they were largely committed in a piecemeal manner trying to plug an increasingly leaking dam. The weakness of the Luftwaffe and Kriegsmarine did not help either.

Remarkably, never once did the Germans waver during the Battle for Normandy, despite losing all strategic initiative in the face of Hitler's stubborn refusal to yield ground until it was too late. Thanks to meticulous planning, the right equipment for the job and a good deal of luck, D-Day was a remarkable success for the Allies on land, air and sea.

16

THE GREAT RIVER CROSSING: THE RHINE 1945

On the night of 23 March 1945, the Allies crossed Germany's last major defensive barrier – the Rhine. After Operation Overlord, this was the second largest operation undertaken by the British Army during the entire war. The Rhineland was the scene of two costly campaigns in the closing months of the conflict. The US 1st Army fought from September 1944 through to February 1945, clearing German forces from the Hürtgen Forest and losing 24,000 men in the process. The Allies then geared themselves up to reach the Rhine itself.

The Rhine is where the Second World War started and finished. The Rhineland held a special place in Hitler's heart as it was where he had started his bloodless expansion of the Nazi regime in 1936, before it ended in bloodshed in Poland three years later. Following the First World War, the Treaty of Versailles had demilitarised the Rhineland, with three occupation zones placed under the auspices of the Allies until 1935, although their forces left five years before the deadline. Despite being intended as a buffer between Belgium, France, Germany and Luxembourg, Hitler's forces reoccupied the

Rhineland in violation of Versailles. Crucially, Britain and France's failure to act at this stage to curtail Hitler's militarism encouraged his subsequent actions against Czechoslovakia and Poland. Hitler half expected the French Army to react and had instructed his troops to turn back at the first sign of trouble, but instead of drawing a line in the sand Britain and France let him have his own way.

Following the defeat of Hitler's surprise Ardennes Offensive in the winter of 1944, the Colmar Pocket unnecessarily distracted Lieutenant General Dwight D. Eisenhower, Allied supreme commander. During late January and early February 1945, the French and American forces attacked General Siegfried Rasp's German 19th Army trapped around Colmar. Although the pocket was sealed by 9 February and the Germans lost 22,000 POWs, the bulk of the 19th Army escaped over the Upper Rhine to fight another day.

In preparation for the advance of Field Marshal Bernard Law Montgomery's 21st Army Group (Canadian 1st, British 2nd and US 9th Armies), the Allied air forces sought to disrupt communications within the German industrial region of the Ruhr and between the Ruhr and the rest of Germany with Operation Bugle. This was followed by Clarion, a major bombing offensive designed to destroy German communications and morale at the end of the month. Significantly, Bugle helped to ensure that much of Field Marshal Walter Model's Army Group B, consisting of General Hasso von Manteuffel's 5th Panzerarmee and General Gustav von Zangen's 15th Army, remained trapped in the Ruhr.

On 8 February 1945, Montgomery launched Operation Veritable, his usual set-piece attack thrusting the Canadian 1st Army under General Henry Crerar in the Netherlands, supported by the British 2nd Army under General Miles Dempsey, into the Rhineland. Attacking through the

Reichwald, Lieutenant General Sir Brian Horrocks' British 30th Corps came up against General Alfred Schlemm's 1st Parachute Army. The German 84th Division was forced out the way, but fierce resistance was encountered from the German 7th Parachute Division. The bitter fighting lasted until 21 February, culminating in the capture of Goch. This was followed by Operation Blockbuster, which took the Canadians to the Rhine itself.

The Americans hoped to launch Operation Grenade across the Roer to the south, but were delayed for two weeks by German flooding. Eleven days after Veritable commenced, Lieutenant General William H. Simpson pushed his US 9th Army forward from Geilenkirchen to the Rhine around Düsseldorf. At the same time, Lieutenant General Alexander M. Patch's US 7th Army advanced to the Upper Rhine.

Hitler was expecting a big Allied push over the Rhine and did everything he could to stiffen the defences of this vast natural barrier. He assessed that the Allies would strike downstream of Emmerich, so Major General Johannes Blaskowitz, commander of Army Group H, deployed the stronger of his two armies there (the 25th, under General Günther Blumentritt). General Schlemm's battered 1st Parachute Army was left to cover the 45 miles between Emmerich and Duisburg.

The German 1st Parachute Army considered itself an elite formation, however its courage was beginning to waver. After the fierce battles in the Reichwald, the *Fallschirmjäger* (paratroopers) were exhausted. On 14 February, the commander of the 7th Parachute Division made his position very clear, 'The sternest measures will be taken against any further unauthorised rearward movements by individual soldiers or small units, of the kind that have been seen during the past two days.'

By 1945 few of the *Fallschirmjäger* had ever made a para-chute drop and their ranks were fleshed out with drafted young recruits or men transferred from the Luftwaffe – none-theless, their morale was surprisingly high. The British had a healthy respect for these German troops in their distinc-tive rimless helmets. 'We felt quite a professional affection for these paratroops,' recalled Corporal Wingfield of the 7th Armoured Division:

> They were infantry-trained like us to use their own ini-tiative. They had the same system of 'trenchmates'. They fought cleanly and treated prisoners, wounded and dead with the same respect they expected from us. If our uni-forms had been the same we would have welcomed them as kindred spirits.

After capturing Xanten on the west bank of the Rhine, the brigadier commanding the 43rd Infantry Division ordered his men to salute the defeated Fallschirmjäger as they filed past.

By mid-February, Montgomery was facing the remains of four parachute, three infantry and two Panzer or Panzergrenadier divisions. It seemed to the Germans that the Allies' resources and firepower were limitless and they knew that continued resistance was increasingly pointless. With the Red Army just 35 miles from Berlin, the continued fighting in the west seemed futile to many senior German officers. Following the German counter-attacks into the Ardennes and Alsace, reserves were scarce. The only mobile reserve consisted of the 47th Panzer Corps with the 116th Panzer and the 15th Panzergrenadier Divisions. They seemed formida-ble, but they could scrape together just thirty-five tanks.

The general prognosis for the *Wehrmacht* did not look good. On 23 March, Montgomery assessed:

The enemy has lost the Rhineland, and with it the flower of at least four armies – the Parachute Army, 5th Panzer Army, 15th Army, and 7th Army; 1st Army, farther to the south, is now being added to the list. In the Rhineland battles, the enemy has lost about 150,000 prisoners, and there are many more to come; his total casualties amount to about 250,000 since 8 February.

While Operation Plunder was the key attack across the Rhine, during the Malta Conference Eisenhower announced additional crossings south of the Ruhr. It was almost as if the Americans were intent on stealing Montgomery's thunder. They launched Operation Lumberjack using Lieutenant Courtney H. Hodge's US 1st Army and General George S. Patton's US 3rd Army, attacking between Koblenz and Cologne on 1 March 1945. The plan was to drive Model's Army Group B back through the Eifel region to the Rhine. Six days later, Hodges met his 7th Corps commander, General Collins, on the Rhine at Cologne. The US 3rd Armored Division drove the remnants of the 9th Panzer Division from the city, but the Hohenzollern Bridge was destroyed.

More importantly, just an hour to the south the armoured Combat Command B of the US 9th Armored Division, supported by elements of the US 78th Infantry Division, famously reached Remagen at the same time. Dramatically, they seized the Ludendorff railway bridge, the only remaining span over the Rhine, before the Germans could destroy it. The Americans had secured a bridgehead, two weeks before Montgomery was ready to go. Ironically, this bridge had originally been constructed during the First World War to move men and materiel to the Western Front. The bridge had two lines and a footpath, but one line had been boarded over to allow road traffic. The Americans wanted Simpson's US

9th Army to cross at Uerdingen, but Montgomery refused, perhaps smarting that he had lost the opportunity to breach the Rhine defences first.

These operations served to distract Hitler southward. Lieutenant General Fritz Bayerlein, commanding the German 53rd Corps, wanted to gather three divisions before counter-attacking at Remagen, but Hitler gave orders for an immediate attack with everything to hand. On 10 March, the newly formed 512th Heavy Tank Destroyer Battalion was thrown at the bridgehead. It was one of only two units equipped with the very heavy Jagdtiger which, although armed with a powerful 128mm gun, was not easy to operate. The 512th were unsuccessful, as were elements of the 9th Panzer Division, and were used to cover the subsequent German withdrawal

Ten days after its capture the battered Ludendorff Bridge fell into the Rhine, killing twenty-eight American soldiers. Its loss mattered little, as by the 21st the Americans had five pontoon bridges over the Rhine at Remagen. In contrast, its capture cost Field Marshal von Rundstedt his job as commander in the west; Albert Kesselring replaced him, but although he was a very able general there was little he could do.

On 8 March the last German pocket of resistance west of the Rhine, at the village of Alpon, was attacked by the British 52nd (Lowland) Division. Troops from the German 1st Parachute Army did not relinquish it easily, in particular, the 6th Cameronians suffered a nasty mauling. Under orders, the defenders finally abandoned the west bank of the Rhine two days later.

While the Allies had to ensure that everything went to plan, the Germans could not understand why the Allied advance was so laborious and on such a broad front. The Allies had substituted Hitler's all-conquering blitzkrieg with a strategy of steady roll-back. Since the failure of Montgomery's single thrust with Operation Market Garden, Eisenhower remained averse to risking his flanks.

Although the Allies assessed that the Wehrmacht was in disarray, they had been very shaken by Hitler's Ardennes Offensive and subsequent tough fighting at Hürtgen and in the Reichwald. In addition, with Montgomery in charge, assaulting the Rhine was destined to be a set-piece battle. Fortunately for the Allies, Montgomery decided to attack upstream of Emmerich where the British 2nd Army faced the German 1st Parachute Army.

The preparations for the crossing took two weeks, which many felt was unnecessary. The Americans were not altogether happy with Montgomery's cautious and meticulous planning. While the Allies were getting ready, the 47th Panzer Corps was able to regroup in the Netherlands and the Germans were able to improve their defences, particularly at Speldrop. For the push across the Rhine, Montgomery wanted the 1st Canadian Army's 2nd Corps and the US 9th Army's 16th Corps to be allocated to the British 2nd Army. The Americans objected and this led to the subsidiary Operation Flashpoint.

Montgomery's 21st Army Group could call on 1.2 million men of Lieutenant General Sir Henry Crerar's 1st Canadian Army (consisting of eight divisions), Lieutenant General Sir Miles Dempsey's British 2nd Army (eleven divisions, including three armoured and two airborne, as well as four armoured, one infantry and one commando brigade) and Lieutenant General William H. Simpson's US 9th Army (eleven divisions, including three armoured).

For ten days before the great crossing, Montgomery's gathering forces at Wesel were shrouded in a dense smokescreen, making it obvious that something big was about to take place. The Allies' logistical effort was enormous, involving 59,000 British and American engineers. The British 2nd Army drew an additional 118,000 tons of stores, including 60,000 tons of ammunition, while the US 9th drew an extra 138,000 tons of stores. Forty-five medium landing craft and a similar number of landing craft vehicle/personnel under Vice Admiral Sir Harold Burrough were shipped to Ostend from the UK. They then made their way to Antwerp under their own power and were collected by army transporters for the onward journey to the Rhine.

As well as the landing craft there were also amphibious DUKWs, Buffaloes and DD Sherman swimming tanks. One of the lead supporting assault formations was Major General Sir Percy Hobart's British specialised 79th Armoured Division affectionately known as the 'Funnies'. This division consisted of five brigades operating almost 2,000 specialised vehicles and gun tanks, including approximately 600 Buffaloes, which were allocated to the various assault divisions.

On 13 March, Patton's 3rd Army crossed the Moselle, then on the night of the 22nd he further stole Montgomery's thunder by throwing the US 5th Infantry Division across the Rhine at Nierstien and Oppenheim, south-west of Frankfurt. As part of the preparations, the US 249th Engineer Battalion was given special training on the floating Bailey bridge in Trier. Captain John K. Addison of the 249th recalled:

On 19 March, our headquarters was at Adenau, Germany. Where we were alerted to join the engineer task force for the Rhine crossing at Oppenheim. We would man the assault boats for the crossing of the second wave of the 5th Infantry Division, to be followed with the construction of the heavy pontoon bridge. Our engineer work went off like clockwork, although one raft was sunk, two of our men were lost, and as many as 200 Germans drowned. The sinking was caused by the sudden shifting of passenger weight brought on by panic.

Hitler immediately declared this a greater threat than the bridgehead at Remagen, as this section of the Rhine was virtually unguarded. Hitler wanted to send a panzer brigade, but all that were available were five disabled Jagdtigers at the tank depot at Sennelager. By the evening of the 24th, Patton had captured 19,000 prisoners.

In the meantime, Montgomery's assault on the Rhine was to consist of two elements: Operation Plunder, covering the British 2nd Army and the US 9th Army crossings at Rees, Wesel and south of the Lippe Canal, and Operation Varsity, involving a massive airborne drop and glider landing near Wesel and the subsidiary Operation Archway conducted by the SAS. They were to take the heights overlooking the crossing at Diersfordterwald and the road and rail bridges over the River Issel at Hamminkeln. This would allow Major General E. Bols' British 6th Airborne Division and Major General William E. Miley's US 17th Airborne Division to prevent the Germans from reinforcing the Rhine with the 47th Panzer Corps.

Varsity, in particular, had to be carefully planned, for it was obvious that the Germans would be expecting an airborne landing as well as river crossings. This time, though, there was to be no repeat of Operation Market Garden, Montgomery's ill-conceived attempt to cross the Rhine at Arnhem. Lieutenant General Lewis H. Brereton, commander of the 1st Airborne Army, in consultation with Montgomery decided to land Major General Mathew B. Ridgeway's 18th Airborne Corps at mid-morning after the crossings had commenced and the ground forces were to link up the same day.

The Allies mighty air fleets also softened up the German defences. During mid-February to mid-March, Bomber Command and the US 8th and 9th Air Forces conducted 16,000 sorties, dropping almost 50,000 tons of bombs. The Germans were attacked at low level from 11–21 March, with the British 2nd Tactical Air Force and the US 24th Tactical Air Command flying 7,300 sorties. The Germans were granted just two days to catch their breath when on the 23 March at 5 p.m. the Allies' artillery bombardment began and continued until 9.45 the following morning.

The formations assigned to spearhead Plunder were the British 30th and 12th Corps and the US 16th Corps. They were to cross at Rees, striking the right flank of the German 2nd Parachute Corps (consisting of the 6th and 8th Parachute Divisions) and its left flank at Xanten, defended by the 7th Parachute Division and the 84th Division (forming the right wing of the German 86th Corps). The crossing at Wesel would also attack the latter's positions. The US 16th Corps crossing at Walsum would attack the German 180th and 'Hamburg' Divisions.

British Prime Winston Churchill, always with one eye on historic moments, could not resist being present. In secret he flew to Montgomery's headquarters near Venlo to watch the beginning of both Plunder and Varsity. On the eve of the crossing, Montgomery signalled, 'Over the Rhine, then, let us go. And good hunting to you all on the other side.'

Montgomery's opening artillery bombardment employed 3,300 guns. At 9 p.m. Horrocks' 30th Corps launched his diversionary attack north of Rees with the 7th Black Watch spearheading the 51st (Highland) Division, followed by the 7th Argyll and Sutherland Highlanders. Horrocks had a good picture of what to expect:

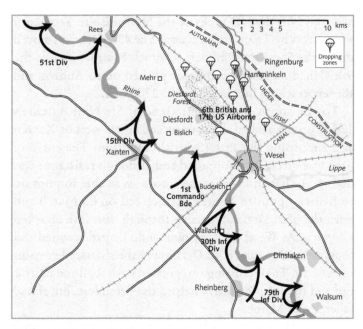

Operation Plunder, 24 March 1945.

According to my intelligence staff whose information was always astonishingly accurate, we were opposed by the 8th Parachute Division round the small town of Rees with part of the 6th and 7th Parachute Divisions on its flanks. Behind in immediate reserve were our old friends, or enemies 15th Panzergrenadier Division and 116th Panzer Division.

The Black Watch took just four minutes to cross and were the first British troops over. The defenders were stunned by the bombing and artillery attacks and offered sporadic mortar and machine-gun fire in response to the crossing. The ground north-east of Rees was quickly secured, but the men of the German 2nd Parachute Corps held onto the town until the next day. All the DD tanks of the Staffordshire Yeomanry were across by 7 a.m. and in action. The 43rd Division moved up on the 51st Division's left to attack Esserden, while the 9th Canadian Infantry Brigade pushed on to Androp and Bienen, reaching Emmerich by the 27 March.

To the south, Lieutenant General Sir Neil Ritchie's 12th Corps launched the main attack north-west of Xanten at 2 a.m. using the 15th (Scottish) Division. The crossing here was virtually unopposed and although resistance was met at Haffen and Mehr, the troops were able to push on to Bislich opposite Xanten. Bislich fell on 25 March and with the 53rd Division passing through, the 15th attacked Dingden. At Wesel the 1st Commando Brigade crossed and fought the German 180th Division in the shattered remains of the city. Their crossing was greeted with shell and mortar fire and the rubble greatly helped the defenders, but it had fallen by 24 March.

The 3rd Canadian, 15th (Scottish), 43rd, 51st (Highland) Divisions and the equipment of the 6th Airborne Division were ferried across by 425 Buffaloes. These made almost 4,000 trips and fifty-five were damaged and nine written off. Special Buffalo troops also laid carpeting to assist the supporting tanks up the very muddy banks of the Rhine. About 100 swimming DD Sherman tanks came with the initial waves, but by 27 March rafts had carried over another 437 tanks. On the far bank the landings were soon greeted by armoured counter-attacks.

Operation Flashpoint formed the right flank assigned to Major General John B. Anderson's US 16th Corps, which was tasked with getting its US 30th Infantry Division across the Rhine between Wesel and Möllen and the US 79th Infantry Division across between Möllen and Walsum. German units consisted of elements of the 180th Division from General

Erich Staube's 86th Corps and the Hamburg Division from General Erich Abraham's 63rd Corps, both of which were part of the German 1st Parachute Army.

South of Wesel, the US 30th Division crossed the Rhine after the German defences were flattened by artillery bombardment and air strikes. By the end of 24 March the US 16th Corps had breached the Rhine defences in five places, securing a bridgehead between Wesel and Dinslaken. This allowed the 8th Armored, 35th Infantry and 75th Infantry Divisions to move up for the drive on Dorsten.

The stunned German defenders were brushed aside and by dawn the Allies' bridgeheads had been firmly established. However, the Germans were quick to recover their wits and the Fallschirmjäger began to fight back. At midnight on 23 March the 15th Panzergrenadier Division was directed towards Rees near the 2nd Parachute Corps sector, while the 116th Panzer Division was ordered across the Lippe to attack the Americans. It found itself having to take control of the front line from the 180th Infantry Division that had disintegrated.

The Germans had anticipated a major airborne drop deep in their rear. They were therefore thrown by Operation Varsity not being simultaneous and barely 5 miles beyond the Rhine. The airborne assault opened at 10 a.m. on 24 March. Now overshadowing the airborne operations in Normandy and the Netherlands, Varsity was the biggest airborne attack of the war. The British 6th Airborne Division committed five British and one Canadian parachute battalion, supported by the air landing brigade with another six battalions of the US 17th Airborne Division.

Lifting 21,700 men required 1,696 aircraft and 1,348 gliders, escorted by 889 fighters. The 6th Airborne flew from Britain and the 17th Airborne came from France. German flak defences were very heavy, consisting of 710 light and 115 heavy guns. Although these were pounded by air strikes and artillery, this did not stop the air armada losing fifty-three aircraft and forty gliders.

Brigadier James Hill's 3rd Parachute Brigade came down near Bergen on the eastern side of the Diersfordterwald, Brigadier Nigel Poett's 5th Parachute Brigade were allocated north-west of Hamminkeln and Brigadier R.H. Bellamy's 6th Air Landing Brigade were south-west of Hamminkeln. Colonel Edson Raff's US 507th Parachute Infantry Regiment (PIR) dropped close to the eastern bank of the Rhine between Bislich and Wesel, the 513th PIR, north-east of Fluren, and the 194th Glider Infantry regiment to the north of Wesel.

The weather was just right and unlike previous air assaults nearly everyone landed on their allocated Drop Zones (DZs). Predictably, resistance was heavy, although the German 84th Infantry Division in the landing zone were caught off guard. The 5th Parachute Brigade suffered mortar attacks before they even hit the ground, but this did not stop them from clearing all the local enemy strongpoints and securing all their objectives within twenty-four hours. The 17th Airborne Corps successfully grabbed all the key points on the River Issel, which runs parallel to the Rhine between Wessel and Emmerich. The German 84th Division all but disappeared and the Germans lost a potential fall-back position.

Although Varsity was a great success, casualties were quite considerable. 6th Airborne suffered 347 killed and 731 wounded, while the US 17th Airborne lost 400 killed and 522 wounded. In total, they captured 3,789 German troops. The 15th (Scottish) Division linked up with them that very afternoon. Subsequently supported by the tanks of the Grenadier Guards and three regiments of artillery, they took just seven days to reach the Baltic port of Wismar, where they linked up with the Red Army.

'The Germans were not going to give up this famous river without a bitter struggle,' recalled General Horrocks, 'and on the evening of the 24th, the 15th Panzergrenadiers launched a vicious counter-attack which was driven back by the 51st Highland Division.' North of Rees, near Speldrop, the Black Watch encountered such heavy German resistance that they had to be rescued by the 9th Canadian Brigade.

On 25 March, Churchill and Montgomery arrived at Eisenhower's headquarters. Following lunch, they went to a house on the Rhine to look at a quiet German sector. After Eisenhower had gone, the pair could not resist crossing with a group of American officers, remaining in enemy territory for 30 minutes. During a later visit to the destroyed Wesel railway bridge they were driven away by German artillery. Churchill was in his element.

At 6 p.m. on the 25th the Americans drove the 116th Panzer Division from Hünxe. The following day the panzers initially found themselves holding the entire 47th Panzer Corps' front until they were assisted by the 180th and 190th Infantry Divisions. To the south 63rd Corps, consisting of the 2nd Parachute and Hamburg Divisions, struggled to hold the line. The latter was made up of staff and communications personnel supported by some *Fallschirmjäger*. During the night of 27–28 March the 116th Panzer withdrew under covering fire from the divisional artillery, and two days later it was issued with just fourteen new *Jagdpanthers*.

South of Koblenz, Patton's US 8th Corps pushed the US 89th and 87th Divisions across the Rhine at Boppard and St Goar at 2 a.m. on 26 March. Altogether, the division plus supporting and attached forces numbered well over 23,000 men. To oppose them, the Germans had Luftwaffe anti-aircraft battalions fighting as infantry and *Volkssturm* home guard equipped with small arms, machine guns, 20mm and 88mm anti-aircraft guns, some field artillery and a few panzers.

The 354th and 353rd Infantry Regiments spearheaded the crossing with the 355th in reserve. The 354th attacked with 1st Battalion towards Wellmich and the 2nd towards St Goarshausen from St Goar. Over a company and a half of 1st Battalion (Companies A and C) reached the east bank on the first wave with little resistance during the crossing, but they came under heavy fire from the hillside behind Wellmich once ashore. Machine gun and 20mm cannon fire, along with the swift current, prevented the assault boats from returning to the west. The 2nd Battalion on the way over was greeted with point-blank grazing fire just above the water from machine guns and anti-aircraft guns. Nonetheless, a pontoon bridge was completed between St Goar and St Goarshausen the following day and over 2,700 prisoners were eventually taken. Further south, Lieutenant General Patch's US 7th Army crossed the Rhine at Worms on 26 March, allowing a breakout towards Darmstadt.

Montgomery now prepared to commit his other four corps, and by 26 March seven 40-ton bridges had been put over the Rhine. This allowed the British 12th Corps' 7th Armoured Division, under Major General L.O. Lyne, and the US 16th Corps' 8th Armored Division, under Brigadier General John M. Devine, to move into the bridgehead. There was no stopping these formations, and by midnight on 28 March the bridgehead had expanded considerably. The 7th Armoured thrust forward 20 miles as far as Borken and the 8th got to Haltern, about 25 miles.

Once the Allies were swarming over the Rhine there was very little Hitler could do to contain them. Within a week of the crossing, Montgomery had amassed twenty divisions with 1,500 tanks and 30,000 German POWs had gone into the 'bag'. In addition, other elements of the British 2nd Army and Canadian 1st Armies were pushing into northern Germany and southern Holland. The US 9th Army was pushing south into the northern end of the Ruhr between Duisburg and Essen.

General Schlemm's 1st Parachute Army was driven north-eastwards towards Hamburg and Bremen, which opened a breach with the German 15th Army defending the Ruhr. Schlemm was wounded and command of his forces fell to General Günther Blumentritt, who along with his superior, Colonel General Johannes Blaskowitz, were under instructions to hold fast at all costs. Both knew that, in reality, this was a pointless exercise. The situation was a shambles, there were few tanks and artillery, and no air cover or reserves to help stem the unrelenting Allied tide. Those reinforcements that did arrive were deemed all but useless, consisting of frightened old men and boys.

The Germans continued to resist, despite it being clear all was now lost. Major Peter Carrington of the Guards Armoured Division was full of praise for his opponents, observing:

> The Germans were very, very good soldiers. After the Rhine crossing, we had 15th Panzergrenadier Division in front of us fighting a rearguard action all the way to the very end of the war; in circumstances in Germany when they must have known they were going to lose the war and didn't have much hope. They fought absolutely magnificently with great courage and skill.

General Bayerlein, commanding 53rd Panzer Corps, was ordered by Model on 29 March to try to break out eastwards with the remains of four divisions. This represented the last major German offensive in the west, but by 2 April he was back where they had started, driven back by American fire-power. Blumentritt, who felt it his duty to save the men under his command rather than throw them away in the defence of the Reich, withdrew behind the Ems Canal towards the cover of the Teutoburg Forest.

Further south, Lieutenant General Courtney H. Hodges' US 1st Army broke out of the Remagen bridgehead, while the spearhead of the US 9th and 1st Armies linked up on 2 April at Lippstadt, east of the Ruhr.

The remnants of Field Marshal Model's Army Group B, some nineteen divisions from the 5th Panzerarmee and 15th Army along with 63rd Corps from the shattered 1st Parachute Army, were caught in the Ruhr pocket. Grandly Hitler dubbed it the 'Ruhr fortress'.

The Allies, in the meantime, pushed on to meet up with the Red Army on the River Elbe. The Ruhr pocket was left to Lieutenant General Leonard T. Gerow's specially created US 15th Army. The outcome was inevitable. Both the 9th and 116th Panzer Divisions surrendered to the Americans in mid-April. On 21 April Major General Joseph Harpe, commanding the 5th Panzerarmee, finally surrendered along with 325,000 men including twenty-nine generals. Model chose death and committed suicide the same day. By this stage, the Second World War in Europe was all but over.

17

A Very Modern Siege: Khe Sanh 1968

The bitter engagement fought in 1968 at Khe Sanh, in South Vietnam, as part of the Tet Offensive, saw tactical airpower coming of age. The American fire-support base at Khe Sanh saw one of the deadliest deluges of munitions unleashed on a tactical target in the history of warfare. The Tet Offensive itself caught General Westmoreland and the American military off guard with an intelligence failure that has been unfairly ranked alongside Pearl Harbor in 1941. Both the siege of Khe Sanh and the battles of the Tet were to have far-reaching ramifications for America's commitment to South Vietnam and the conduct of the war.

Formerly part of French Indo-China, Vietnam was partitioned by the Demilitarised Zone (DMZ) in 1954 between the Communist north and the non-Communist south. Although the French were badly defeated at Dien Bien Phu, they could have continued the war. However, General Vo Nguyen Giap's victory was such that France lost the will to continue the struggle against communism in Southeast Asia.

Between 1950–54 America had provided $1.1 billion to help France with the war against communism in Indo-China, including $746 million worth of equipment delivered direct to the French Expeditionary Force (FEF). It was not enough to help stem the tide. The Battle of Dien Bien Phu had opened with a French paratroop drop on the Vietminh-held village in 1953. It was one of three sites that the French hoped to use to pin down the Communist interior. The garrison of 6,500 men, comprising French, Algerians, Moroccans, Thais, Vietnamese and Montagnards found itself besieged by 50,000 Communist troops.

The defenders had to endure a bombardment of about 150,000 shells during March–May 1954. All their supplies had to be flown in from Hanoi, 300km away. They held out for a grim eight weeks until finally forced to surrender. The humiliated French suffered 2,000 dead, 7,000 wounded and missing and 7,000 POWs, at a cost of 8,000 dead and 15,000 wounded Communists. Air support had been insufficient; the garrison had been denied reinforcements and prevented from breaking out. The disaster had all the hallmarks of Stalingrad in 1943 and it was a lesson the North Vietnamese and Americans were not to forget.

By 1955 the FEF had been reduced to 35,000, with the withdrawal accelerating the following year – the French had much more pressing matters in Algeria. Determined to see the two halves of the country united, the South's Vietcong, or Vietnamese Communists (VC), began a terrorist campaign in the late 1950s. Then in 1960 the VC were incorporated into the National Liberation Front (NLF), which called for a sustained guerrilla war rather than a North Korean style conventional war to forcibly unite the country.

The Army of South Vietnam consisted of a collection of former colonial troops. With American help, it was slowly moulded, gaining some counter-insurgency capabilities, but political upheaval undermined it. The Buddhist uprising and the fall of the government in 1963 led to the deterioration of the South Vietnamese Armed Forces (SVAF) and only the intervention of US combat forces saved the Republic.

American military advisors, who became designated the US Military Advisory Command Vietnam (MACV), began to arrive in the early 1960s with the first combat troops arriving in 1965, followed by Australian and New Zealand forces. South Korea and Thailand were also to provide men. The North responded by committing regulars of the North Vietnamese Army (NVA).

America's strategy was to set up fortified bases along the coast (Phu Bai, Da Nang, Chui Lai, Qui Nhon and Cam Rahn) from which to conduct operations. There were to be no front lines, except round the main bases, meaning the countryside was left to the mercy of the VC. The Americans' primary goal was to cut the Ho Chi Minh Trail, down which the North supplied the Communist forces operating in the South. Airpower was chosen as the instrument to achieve this.

Between 1965–68 American forces were to bear the brunt of the fighting. In 1954–55 the total number of SVAF stood at 279,000, but by 1968 it had almost tripled to 820,000, while the Army of the Republic of Vietnam (ARVN) had grown from 170,000 to 380,000. However, in 1967 the ARVN had been relegated to a secondary role, pacification and security operations. The large-scale US involvement precluded equipment modernisation for the ARVN until after 1968. This was to have far-reaching ramifications for the US military.

Furthermore, absenteeism and desertion from the ARVN was a problem and large numbers of men were missing at the start of Tet. The rate rose from 10.5 per 1,000 to 16.5 in July, amounting to some 13,000 desertions, the highest monthly rate since 1966, reaching an all-time high in October. The apathetic attitude ended sharply with Tet and clearly the

ARVN should have been given a greater share of the fighting between 1965–68. The general condition of the SVAF can only have encouraged the Communists.

By the end of 1967 publicly America was winning the war, having declared NVA/VC losses at 167,000 men (88,000 killed in action, 30,000 dead or disabled from wounds, 6,000 POWs, 18,000 defectors and 25,000 lost to desertion and disease), but on the ground things were slightly different. Official reports of NVA/VC strength in South Vietnam were double those released to the public.

General Giap felt that the time was ripe to repeat the Dien Bien Phu strategy and deliver a body blow to America's resolution to prop up the South. Divisions in the American Government and the growing peace movement only served to encourage him. US intelligence by late 1967 indicated that four NVA infantry divisions with two artillery regiments and supporting armour, totalling 40,000 men, were moving on the Khe Sanh area in the northern most Quang Tri Province. Notably, the Tet Lunar Truce was not to encompass I Corps, whose area of responsibility included the five northern provinces of South Vietnam. Infiltration of the South was stepped up and by the end of 1967 the NVA/VC had more than 60,000 men in South Vietnam.

The NVA/VC Tet Offensive effectively started in September 1967 when the Communist forces launched attacks against the isolated American garrisons in the central highlands. They stepped up their activity by early January 1968 with an attack on Da Nang air base, destroying twenty-seven aircraft. Attacks were also made in the Que Son valley and against the Ban Me Thuot airfield and An Khe. The VC announced a seven-day truce for the Tet holiday, which was schedule to begin on 28 January, and VC activity died down the day before apart from shelling of the base at Khe Sanh. Nevertheless, the South Vietnamese Government stated the

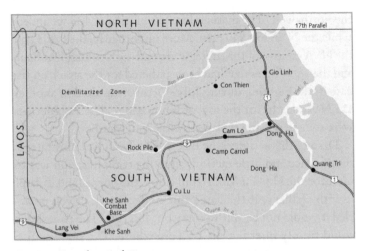

Vietnam Demilitarised Zone.

truce would not be observed in the five northern provinces due to the VC build-up during the month.

Khe Sanh, a few miles from the Laotian border in the western half of Quang Tri Province, had been a Special Forces camp until 3,500 US Marines arrived in 1967. The Green Berets then moved to the Montagnard village of Lang Vei, closer to the Laotian border to the south-west. Khe Sanh, straddling Route 9, an old French road linking Dong Ha on the coast with Laotian towns along the Mekong, was then expanded to act as a springboard for proposed operations against Communist bases in Laos and the bordering DMZ. Khe Sanh's defences were built around the vital airstrip and three hills to the north-west, 881 North, 881 South and 861A, covering Route 9. A detachment was also on Hill 950 to the north, on the other side of the Rao Quan River, while to the south lay the Special Forces camp near Lang Vei.

The US 3rd Marine Regiment had secured the hills during April–May 1967 after prematurely triggering an attack from the north-west. The marines lost 155 killed and 425 wounded, while the NVA lost 940 men. More importantly, the battle cost the NVA control of the strategic ground overlooking Khe Sanh and the very forces that had been gathered for the attack on the base itself. Fighting during July–October resulted in another 113 NVA deaths for the loss of ten marines. The 3rd Marines were rotated out and replaced by the 26th Marines.

Despite this activity, the base at Khe Sanh had no strategic significance and little tactical value. Its main function was to monitor the Ho Chi Minh Trail across the border and disrupt it with artillery. If the base came under attack or was encircled it would not be able to carry out this role. The Americans could simply interdict the trail elsewhere. The site was also difficult to defend, as its security relied on holding the three hills and the water supply passed through enemy territory.

The marines wanted to abandon the base but General William C. Westmoreland, commander MACV, was of a different opinion. He no doubt felt it would be unwise for American troops to give ground in the face of a conventional rather than guerrilla NVA/VC attack. Westmoreland wanted to fight a conventional battle and was adamant there would be no repeat of Dien Bien Phu. Furthermore, as enemy action was expected over the 1968 Tet holiday the American command felt it prudent to counter it in an area where it was most anticipated, thereby tying down Communist forces and preventing them from causing mischief elsewhere. Whatever the case, the marines were ordered to stand and fight largely for symbolic reasons. This decision also left the isolated garrison at Lang Vei to its fate.

The base was of greater significance to the NVA/VC placed in the north-east corner of South Vietnam, just 14 miles from the DMZ and 6 miles from the Laotian section of the Ho Chi Minh Trail. Its shelling of the trail was a major nuisance and its capture would ease logistic problems and be a morale booster. The Communists would be able to easily sustain an attack and their long-range artillery could operate from the safety of the DMZ. Ironically, Giap's strategic plans ran on similar lines to the MACV's but in reverse. Khe Sanh was to direct attention away from a much wider enterprise, which the Communists were planning for the whole of South Vietnam.

Just three days after America called a halt to the bombing of Hanoi and six days after the halt to bombing of Haiphong (North Vietnam's main port) in preparation for the Tet Holiday Truce, the North struck at Khe Sanh. They allowed themselves only ten days before the more general Tet Offensive, so they had to act quickly. General Giap's strategy was to take the surrounding hills and then use them as artillery positions to seal the airstrip. The American forces at Lang Vei were to be cut off and destroyed while Khe Sanh would first be worn down and then overrun. Giap's major failing was to underestimate America's ability to resupply their forces by air: in effect, it would prove almost impossible to cut the garrison off completely.

A marine patrol bumped into NVA forces on 17 January 1968 south-west of Hill 881N. Over the next few days, contact with the enemy continued. A prisoner from the North Vietnamese 325C Division informed the marines that the NVA were planning attacks on Hills 881S and 861. This was to be followed by a thrust from the east, with diversionary attacks from the northern and western sector. The main attack was to be conducted by the 304th Division which, using the woods as cover, was to seize the airstrip. Both sides knew that without the airstrip resistance would crumble. Late on 20 January, 300 NVA launched an attack on 861 as predicted, but their reserves were pinned down by artillery fire and the assault repulsed. The attack on 881S never materialised.

In the early hours of 21 January, Khe Sanh base received its first direct attack; in the shelling eighteen Americans were killed and forty wounded, the main supply dump was destroyed with 1,300 tonnes of precious ammunition, a helicopter was wrecked and five others were damaged. Probing attacks were also launched against the marine garrisons holding the three hills and against Lang Vei. At Khe Sanh village the defenders suffered two attacks by the 304th Division, the next day they were forced back into the base compound. The marine commander at Khe Sanh base, Colonel David Lownds, had some of the villagers and local Vietnamese civilians flown to Da Nang and Quang Tri. Nevertheless, about 6,000 refugees had to be excluded from the base during the course of the battle. He also tried to keep the road open to Lang Vei.

The MACV's response was swift – a supporting air campaign dubbed Operation Niagara was implemented. Khe

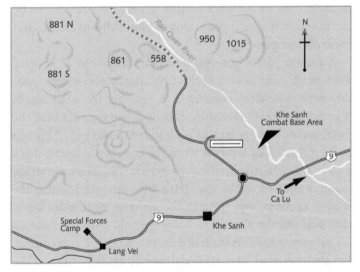

Khe Sanh Valley.

Sanh received 1,500 reinforcements, boosting the garrison to over 5,000 and USAF planes and army helicopters flew in supplies under fire on 22 January. A fresh battalion of marines was deployed and some two-thirds of the infantry were placed on the hills, including about thirty men on Hill 950 and about 400 men on Hill 558. The rest were used for perimeter defence of the base itself, supported by artillery, mortars and five M48 tanks.

It was estimated that 18,000 Communist troops had been gathered for the attack, including the 304th and 325C NVA divisions (which had gained fame thirteen years earlier at Dien Bien Phu – the symbolism of this was not be lost on those involved) with twenty-seven tanks and 122mm heavy field artillery. Another division, the 320th near Camp Carrol, 15 miles north-east, was positioned at Rock Pile, cutting Route 9. This obstructed American military movements from Quang Tri and Dong Ha. The 324th NVA Division was also available for the coming siege.

To make matters worse, the weather changed enveloping western Quang Tri Province in thick fog. Under cover of this, the NVA launched a bombardment on 24 January using 152mm artillery: the Americans suffered seven dead and seventy-seven wounded. The attackers also dug positions close enough to hit the indispensable airstrip with mortar fire. When the weather lifted on 26 January, all US air operations in the whole of South Vietnam were directed in support of the besieged base, with 450 dedicated sorties that day. The elite 37th ARVN Rangers Battalion was flown in and these 318 men were to fight with distinction alongside the US Marines. Forces also attempted to reopen Route 9 resulting in 150 VC and nineteen American dead.

Estimates of NVA strength had by now reached 50,000 men: the 324B Division was near Con Thien and Gio Linh, bringing Communist forces up to 40,000 infantry and 10,000 gunners, engineers and other support troops. Khe Sanh's garrison of 6,000 men, consisting of the 9th and 26th US Marine regiments, supporting artillery and the ARVN Ranger battalion, were tying down over 20,000 NVA in the immediate vicinity. This was to have consequences for NVA/VC operations elsewhere.

Reports seemed to indicate that the Communists had committed everything to the battle, but General Westmoreland was rightly suspicious of the lull in the rest of South Vietnam and anticipated diversionary actions. Planning for the relief of Khe Sanh began only four days after the battle started, but had to be postponed. Indeed, Khe Sanh was quickly forgotten on 30 January 1968 when the NVA launched their Tet Offensive with over 80,000 men. In fact, some attacks were launched twenty-four hours prematurely, tipping the Americans off that something big was about to happen. Within five days, thirty of South Vietnam's forty-four provincial capitals, including Hue, had fallen to them. Tet was intended to be the main Communist offensive to conquer large parts of South Vietnam and combined with an American defeat at Khe Sanh would, felt the North's leadership, bring a swift end to the war.

Giap's grand strategy was to spread the US/ARVN forces to a point where they would be unable to counter his primary goal. Despite the widespread attacks, two areas emerged as major targets – the cities of Saigon and Hue. These were to be the only places where the fighting was protracted. Some 5,000 NVA/VC troops had successfully infiltrated Saigon and launched their surprise attack at 3 a.m. on 30 January. Over 700 men attacked Tan Son Nhut Airfield, outside Saigon, and the neighbouring MACV compound to eliminate the command post of the US 7th Air Force.

Attacks were also staged against the presidential palace, the South Vietnamese joint general staff headquarters, the radio station, US Navy headquarters, the Philippine Embassy and the American Embassy. The suburb of Cholon was occupied and a headquarters set up in An Quang Pagoda. The heavy fighting around the MACV compound forced Westmoreland to withdraw to his command bunker and order his staff to draw weapons to assist the defenders. The following day, martial law was declared and a curfew imposed. It took until 5 February to stabilise the situation and confine resistance to Cholon.

The NVA's real success was at Hue where the 4th and 6th NVA regiments with six VC battalions captured most of the city and held it for a month. The attack on Hue was launched on the same day as Saigon and by nightfall the NVA/VC had raised their flag over the ancient citadel. During the day, US/ARVN forces pushed into the city to rescue some American advisors and then promptly withdrew. Only in Hue did the attackers receive any measure of support, principally from the student population. The commitment of US/ARVN troops to Khe Sanh and other locations in the northern province had the desired effect, for there was a delay in gathering sufficient assault forces to regain the initiative in Hue.

During Tet, the Communists did not let up their pressure on Khe Sanh, though the expected all-out assault never took place. Technology took a further hand in the base's defence in the form of acoustic and seismic sensors, of which 250 were placed around the perimeter over a ten-day period. On 4 February it was reported that there were five NVA divisions south of the DMZ, now that the 308th Division had been brought into the area. This eavesdropping forewarned the marines that the Communists were gathering near Hill 881 with a view to attacking it on 5 February.

However, the main assault was actually launched against Hill 861A, catching the defenders off guard and allowing the NVA/VC to almost overrun them. Recovering from the surprise, Lieutenant Donald Shanley led his men in a counter-attack and for over 30 minutes a desperate hand-to-hand battle was fought. It was not until the barrage around Hill 881 had dispersed the diversionary attack and been brought to bear on 881A that the NVA/VC were driven off. Due to the terrain, the Communists dug only one tunnel and this was near Hill 861A. It was detected and destroyed.

The NVA made little use of its armoured forces; a notable exception to this was the attack on Lang Vei Special Forces camp when they deployed thirteen PT-76 light tanks. The garrison, under Captain Frank Willoughby, consisted of a Special Forces 'A' Team of twenty-four Green Berets, four under-strength Civilian Irregular Defense Group companies of 400 Montagnards, a detachment of Vietnamese Special Forces and 500 Laotian soldiers (who had been driven over the border by the NVA inside Laos). They were equipped with two 106mm and four 57mm recoilless rifles, two 4.2in and six 81mm mortars, as well as M-72 light anti-tank weapons (LAWs). The marine artillery at Khe Sanh to the east could be called on, as well as AC-47 Spooky gunships in case of night attack, F-4 Phantoms and Huey helicopter gunships. Unfortunately, an MACV directive dismissed the possibility of tank attack so the camp crucially had no anti-tank mines.

The NVA launched their attack at 10.30 p.m. on 6 February 1968 using elements of the 304th Division. Their plan was to occupy the base and ambush any relieving force. They quickly breached the wire and were in the forward trenches with the Montagnards. From the observer tower on top of the camp's Tactical Operation Centre (TOC), two PT-76s were spotted, five more were also seen approaching another position. Calls to Khe Sanh and Da Nang informing that enemy tanks were in the wire were met with initial disbelief. Communication with Khe Sanh went as follows:

'Intrigue, Intrigue, this is Brass Study, over! We are taking a heavy ground attack and have armour in the wire. Stand by for fire mission over.'

Khe Sanh came back, 'Brassy, this is Intrigue. Are you sure about the armour?'

'Roger, roger, that is affirm. We have tanks in the perimeter.'

'Can you see them from your location?'

'Affirmative, affirmative! I can hear the engines back-firing. They're firing into the bunkers!'

'Negative, Brassy. That must be the sound of your generators backfiring.'

'Intrigue, be advised one of our generators just blew down the bunker door!'

The defenders disabled three or four tanks and one was knocked out by an American aircraft (the first recorded tank kill by a helicopter did not occur until 1973 at An Loc), but they were still overwhelmed at the perimeter.

The commander at Khe Sanh was reluctant to send helicopters or ground forces to a hot site at night – Route 9 would be blocked and NVA tanks were on the landing zones. Instead, artillery at Khe Sanh began to fire onto the perimeter and at 1 a.m. F-4s and A-1 Skyraiders launched air strikes. A B-57 Canberra bomber triggered fifteen secondary blasts which may have damaged three more of the NVA tanks. In the darkness, and with nothing to effectively counter the remaining tanks, organised resistance collapsed. The M-72 LAWs proved defective and the surviving defenders retreated to the TOC where they were trapped until 4 p.m. the following day. Friendly fire was called down on the TOC and in the confusion Willoughby, thirteen Green Berets and sixty Montagnards escaped to be airlifted to Khe Sanh.

At the same time, about 500 men of the 325C Division attacked thirty marines defending an outpost near Rock Quarry. The marines were driven from their positions and their officer killed. The following day, with massive fire support, it took the marines just fifteen minutes to recapture the outpost.

As Lang Vei was falling, the Communists launched another attack at Khe Sanh on 8 February, against the positions of the 9th Marines. The NVA/VC achieved half their objectives before massive air, artillery and tank support stopped them in their tracks. The huge B-52 bombers even dropped bombs to within 100m of the base's perimeter. Under the cover of the air strikes and M48 Patton tanks, a relief force drove off the attackers leaving 150 dead.

For five days the NVA/VC were pounded from the air. Then the weather closed in, enabling them to complete their elaborate siege network of tunnels, trenches and bunkers surrounding Khe Sanh. More importantly, anti-aircraft artillery on the surrounding hills was proving effective against American cargo planes, forcing them to rely on helicopters. Due to losses, the US 7th Air Force banned the C-130 Hercules from operating at Khe Sanh, leaving air supply to the C-123 Provider. The loss of one of these and forty-eight lives on 6 March resulted in increasing reliance on helicopters.

Crucially, Giap had underestimated the technological strides that had been made since Dien Bien Phu. The helicopter sustained the lifeline to Khe Sanh. Indeed, air supply delivered 12,000 tons of cargo with the loss of six fixed-wing aircraft and seventeen helicopters. In addition, over 600 para-drops were made, which were vital for those men on the exposed hills. On the ground, the Communists did ensure that no support got through by keeping up their artillery attacks on American bases at Dong Ha, Con Thien and Camp Carroll.

Further afield at Saigon, the American and ARVN attempted to dislodge the Communists from their stronghold in the Cholon suburb, where they had some 1,600 men. NVA/VC forces launched a counter-offensive from Cholon and other parts of the city on 18 February. Tan Son Nhut was attacked again but the NVA/VC got no further than the outer perimeter. Reports estimated that up to three NVA divisions were outside the city, although they failed to get in and the central area had been pacified three days later.

At Hue the marines were ready by early February after Lieutenant General Cushman, I Corps Marine commander, and Lieutenant General Lam, Vietnamese force commander, had rounded up enough men. Three US Marine and eleven ARVN battalions fought a bloody two-week battle for Hue. The assault on the citadel did not commence until 20 February. Five days later, the city whole was described as secure. The Communist forces lost 8,000 men in Hue and its surrounding area. Although the Tet Offensive was over by 26 February, Khe Sanh remained under siege.

Despite the failure of Tet, on 29 February a large-scale night attack was launched against Khe Sanh. Eight days earlier, about 100 NVA had probed the lines. The sensor technology again warned the defenders of the build-up at the eastern end of the main perimeter. Artillery, tactical aircraft and B-52 bombers ensured the attack died out before it reached the ARVN rangers' position. During March, 325C Division to the north and 304th Division to the south squeezed the base. Elements of the two divisions launched another attack on 17 March with the view to destroying the perimeter defences. The following day, an NVA battalion tried to exploit the weakened sector and persisted for two hours before the marines repulsed them.

Probing attacks continued for several days then, perhaps signalling a weakening of Communist resolve, they switched to artillery bombardment. At the end of March, the besieging forces had fallen by 2,000 to a total of about 10,000. The headquarters of 325C Division had withdrawn to Laos. Between 23–24 March the artillery fire was such that the marines were confined to their bunkers. In two days, the base endured some 1,265 incoming rounds. The defenders, however, remained active, launching a raid against an NVA battalion position less than a kilometre from the base, killing 115. A total of 1,602 NVA bodies were found in the immediate area of Khe Sanh, although total enemy losses were estimated at 10,000–15,000.

By now, time was running out for the attackers, Tet was at an end and the situation in South Vietnam stabilised. The Americans and ARVN had been putting together a relief force, numbering 30,000 troops, for Operation Pegasus/ Lam Son 207. The 1st Cavalry Division (Air Mobile) landed 10 miles from Khe Sanh on 1 April, they joined other American and ARVN forces to open Route 9 so that supplies could get through. The 9th Marines from Khe Sanh attacked and captured an enemy hill dominating Route 9 on 4 April, they then drove off an enemy counter-attack the following day. The advance of the relief force was slowed not by the NVA but by mines and fifteen bridges in varying states of repair. Finally, elements of the Air Mobile division entered Khe Sanh largely unopposed on 7 April. Fighting was to continue, but in effect the seventy-seven-day siege was over.

A US/ARVN force occupied Lang Vei on 10 April and a seesaw battle for its possession lasted until 17 April when the Americans finally abandoned it. On 15 April, the NVA resumed its troublesome bombardment of Khe Sanh. The 304th NVA Division, recovering from its treatment, triggered the battle of Foxtrot Ridge after attacks on Route 9, west of Khe Sanh. To protect the road, 100 marines were placed on the nearby ridge and in the early hours of 28 May an NVA battalion attacked them. For nearly nine hours the NVA pressed home their attacks but were finally driven off with 230 dead, while the marines suffered thirteen killed and forty-four wounded. Two subsequent battles resulted in another 140 NVA dead.

Military activity in the area continued well into the following year. Operation Scotland II began on 15 April, ran until February 1969 and resulted in over 3,000 enemy casualties around Khe Sanh. The base itself was evacuated in mid-1968 and was temporarily recaptured by the ARVN in 1971. After that, it remained in NVA hands and by 1973 the airstrip had been turned into an all-weather MiG fighter runway.

Tactically, Khe Sanh had been a success for North Vietnam; strategically, however, Tet was a failure. The assault on Khe Sanh and the hinterlands before and during Tet were designed to draw American forces away from the South's urban areas. Although the siege was just a feint, the Americans, instead of withdrawing, took up the gauntlet. General Westmoreland saw the situation in reverse, thinking that the attack on the cities was an effort to divert attention from Khe Sanh. He had expected the enemy's effort would be directed at South Vietnam's northern provinces and his men were deployed accordingly. In many ways, the NVA divisions committed to Khe Sanh would have been better utilised in the battle for the cities, particularly Hue. Both the Americans and Vietnamese seemed to hope that Khe Sanh would become a Dien Bien Phu.

Nonetheless, the Dien Bien Phu comparison is not really fair. The French Army had been caught in a valley with little artillery support, while the Americans had extensive fire support inside and outside the Khe Sanh perimeter. American artillery fired almost 160,000 rounds during the siege. The French had also lacked air support, while the US was able to call on helicopters and transport aircraft for supplies as well as bombers to deliver heavy ordnance. For example, over 100 million pounds of napalm had been dropped in the surrounding area by 15 February.

Naval air strikes were also used in support of the marine defenders in early 1968; during February and March, 3,100 sorties were flown by carrier-based aircraft alone. US air support for Khe Sanh also included over 7,000 sorties by the 1st Marine Aircraft Wing. In total, over 24,000 tactical sorties and 2,500 B-52 sorties were launched against the surrounding Communist forces. Over 100,000 tons of bombs were expended. In particular, the B-52s, dropping 75,000 tonnes of munitions over nine weeks, provided the heaviest firepower ever unloaded on a tactical target. Giap himself almost became a victim of the B-52s when he visited Khe Sanh in late January. Operation Niagara had even initially included consideration of tactical nuclear weapons.

The ratio of casualties also highlights a difference with Dien Bien Phu. In 1954, losses were 8,000 Communists to 2,000 Frenchmen; in 1968, they were 10,000 Communists to 500 Americans. According to the North Vietnamese, up to 90 per cent of the NVA/VC losses were due to bombing and artillery fire. Tet, in general, was a disaster for the NVA/VC, more than half their 80,000 men involved were killed. The net result was that it would take another seven years before the North would secure its long sought-after victory over the South.

The North's major failure was to inspire the population of the South to rise up against its government and the Americans. It was then only a matter of time before the NVA/VC forces were overwhelmed. Giap blamed poor communications and co-ordination. The NVA/VC also underestimated the improved quality of the ARVN, who bore the brunt of the initial attacks.

The Tet Offensive resulted in so much damage and destruction that it was hard to see any real winner. America declared

Khe Sanh and Tet an overwhelming victory, but in terms of victor, it was really a draw. The American and ARVN forces won on the battlefield, but the offensive highlighted the extensive reach and capabilities of the NVA/VC. Although Giap's military was severely mauled, he did achieve a Dien Bien Phu goal in that world opinion was increasingly against the war and the American public were questioning its legitimacy. To the public, it graphically illustrated what a tenuous hold the MACV had on the situation. In fact, Tet and Khe Sanh proved to be the foundation of the American withdrawal from the Vietnamese imbroglio. The Tet Offensive was a major reason President Johnson withdrew his candidacy for re-election, thereby forcing an administration change on the US, which in the long run could only be good for North Vietnam. Johnson's greatest failing was not taking a firmer stand after Tet.

General Westmoreland returned to America to become chief of staff. Peace talks commenced in Paris in May 1968 and the following year America began to reduce its military presence. After Tet, there was a general mobilisation in South Vietnam, heralding a policy of 'Vietnamisation' – 1968–72 saw a major expansion of South Vietnam's armed forces.

The North Vietnamese became more conservative during 1969 because of their losses and because of the Americans' process of Vietnamisation. The North was biding its time, infiltrating 115,000 men into the South to make good its losses. The bulk of the American, Australian, New Zealand, South Korean and Thai troops pulled out in the early 1970s. Despite the peace treaty of 1973, the North kept up the military pressure and the ARVN disintegrated in 1975, leaving the South finally in Communist hands.

18

TRIUMPH IN THE AIR: BEKAA VALLEY 1982

In June 1982 in eastern Lebanon some 200 jets fought one of the biggest aerial battles in history, high above the Bekaa Valley. During the engagement, the Israeli Air Force destroyed a considerable Syrian air defence system and inflicted a defeat of the first magnitude on the outclassed Syrian Air Force. No operation in the history of air warfare stands comparison, except that of the 1967 Israeli attack on the Arab air forces. However, then the enemy was caught largely unprepared on the ground. Not since the Second World War have so many aircraft been lost in a single airborne action, nor have such losses been sustained entirely by one side.

Syria always regarded Lebanon as part of Greater Syria and constantly sought to manipulate Lebanese politics. Lebanon achieved independence from France in 1943. Just five years later, with the creation of the State of Israel, 400,000 Palestinian refugees moved north from Israel and crossed into southern Lebanon. For twenty years they lived in refugee camps dreaming of an independent state of Palestine. Then, in 1968 Beirut became the headquarters of the Palestinian

Liberation Organisation (PLO), the umbrella organisation for the different political factions dedicated to the overthrow of Israel. The PLO set up bases in southern Lebanon to carry out cross-border operations in Israel. Largely unwillingly, the Lebanese were dragged into the Arab–Israeli conflict.

Lebanon was then split along fractional lines during the civil war of 1975–76, resulting in 80,000 deaths. The PLO was unable to keep out of this internecine warfare and was soon fighting Lebanese Christian militias. In 1975, at the height of the civil war, Syria despatched the Saiga (a Syrian-sponsored PLO faction) and the Palestinian Liberation Army to fight alongside the Muslim forces. In turn, the Christians supported the Syrian-Lebanese peace plan, but the PLO refused to co-operate. Syria sent in 40,000 troops sanctioned by the Arab League, to help the Christians and prevent the PLO establishing a state within Lebanon capable of resisting Syrian aspirations. The Syrians consolidated their positions in Beirut and the Bekaa Valley. After the 1979 Camp David Accord between Israel and Egypt, Syria's relations with the PLO improved and the Syrians withdrew from the coast. Despite this, by 1981 Syria still controlled two-thirds of Lebanon.

Harassed by Christian forces, the Syrian Army laid siege to Zahlé, a Greek-Catholic city in the Bekaa in April 1981. It was the third largest in Lebanon, with a population of 200,000 Greek Orthodox Christians. More importantly, it was the capital of the Bekaa Valley, an area Syria regarded as vital to the defence of Damascus. The Israelis, in support of the defenders, shot down two Syrian helicopters. In response, Syria began to construct emplacements for batteries of Soviet-built SA-6 surface-to-air missiles (SAMs) in the Bekaa. Although Israel immediately planned to remove them, crucially they were still in place in 1982. Israel's primary concern in helping the Lebanese Christians was to secure its northern border against the frequent PLO incursions. With the PLO operating against northern Israel from havens in Lebanon, it was clear that Israel and Syria were likely to come into direct conflict over the prostrate country.

Israeli planning for the invasion of Lebanon had started in November 1981. The Israeli Defence Force (IDF) intended to deploy some six and a half divisions, with about 75,000 men, 1,240 tanks and 1,500 armoured personnel carriers. Forces were also kept on the Golan Heights to deter the Syrians from attacking the base of the IDF advance. To resist invasion, the PLO had three conventional formations (each of 2,000–2,500): the Yarmouk Brigade stationed along the coastal strip; the Kastel Brigade in the south; and the Karameh Brigade on the eastern slopes of Mt Hermon, in an area the Israelis called Fatahland. The PLO had approximately 15,000 men altogether, with 6,000 in Beirut, Ba'abda and Damour, 1,500 in Sidon, 1,000 between Sidon and Tyre, 1,500 in Tyre, 1,000 stretching from Nabatiyeh to Beaufort Castle, 2,000 in Fatahland and 1,000 in the UNIFIL Zone in the very south. These forces were equipped with about eighty artillery, heavy mortars and rocket launchers.

The Syrian Army also had about 30,000 men and 712 tanks deployed in two main areas of Lebanon. First, in the strategic Bekaa Valley (1st Armoured Division, comprising the 91st and 76th Tank Brigades, plus the 58th Mechanised Brigade with the 62nd Independent Brigade attached and nineteen SAM batteries). Second, in the Shouf Mountains and additionally in the Beirut area (the 85th Infantry Brigade with twenty commando battalions) to guard the important and vulnerable Beirut–Damascus Highway.

Significantly, the Syrian Air Force (SAF) had upgraded its equipment during the early 1980s. It had some

90,000 personnel, equipped with 650 combat aircraft, including 130 MiG-23 and Sukhoi Su-22 fighter-bombers, with another 300 older MiG-21s and Sukhoi Su-7s. The number of assault helicopters had doubled, with the emphasis on attack and anti-tank types, such as the French Gazelle and the Soviet Mi-24 Hind. Notably, Syria purchased twelve Hind attack helicopters in 1981 and by the following year had forty-five Gazelles, mainly in the anti-tank role.

Perhaps the greatest cause of concern for Israel was the fact that the Syrian air defence forces had expanded considerably. For example, anti-aircraft batteries had increased from thirty-four to 100. In attack or defence, the Syrians felt from previous experience that the solution to Israeli tactics was saturation air defence of the combat zone by SAMs, anti-aircraft artillery (AAA) and interceptor aircraft. At least five different SAM systems were to be deployed in Lebanon to provide an impenetrable airspace, consisting of the SA-2 Guideline and SA-3 Goa, used in conjunction with the former SA-6 Gainful self-propelled triple launcher (Syria had thirty batteries of these), SA-8 Gecko and SA-9 Gaskin. Most of them were radar guided.

Heyl Ha'Avir, the Israeli Air Force (IAF), under Major General David Ivri, numbered 30,000 men equipped with 600 combat aircraft, including A-4 Skyhawks, F-4 Phantoms, F-15 Eagles, F-16 Falcons and Kfir C-2s. Despite its age, the A-4 Skyhawk was still an effective ground-attack plane and Israel had about 150 of them. The Kfir C-2 is an Israeli redesign of the French Dassault Mirage V and was Israel's main close-air-support plane. It was up to these forces to destroy the Syrian air defences in the Bekaa if necessary.

The Syrians had every reason to be cautious of the IAF and had a grudging respect for its capabilities. In 1967 during the Arab–Israeli Six Day War, they lost sixty aircraft in a single day to Israeli air strikes. After securing air superiority, the outcome was never in doubt for Israel. In just twenty-five minutes the Israelis destroyed the Jordanian Air Force, damaged the Iraqi Air Force and inflicted such losses on the SAF that it took no further part in the fighting. The IAF's timing and co-ordination was superb, particularly in catching the Egyptian Air Force as the early morning mists lifted on the Nile Delta. At the end of the first day, 300 Egyptian planes had been destroyed and for the whole of the conflict Israel claimed 418 Arab aircraft for the loss of twenty-seven. Five years later in 1971, Pakistan tried to emulate Israel's success with an air attack on India. It was conducted on too narrow a front with insufficient depth, the result was that Pakistan claimed eighty-one aircraft while India claimed ninety-four.

Israel's overall strategy was to advance up the Lebanese coast, bypassing the major PLO centres and cutting off their escape route through the Bekaa. Only then could the PLO be destroyed piecemeal. On the central front, the old Crusader stronghold of Beaufort Castle was an important initial objective. Overlooking southern Lebanon and northern Israel, it had been a PLO artillery observation post for years.

Fortunately for Israel, global attention was focused elsewhere. On 26 May 1982, American Secretary of State Alexander Haig stated, as Iran began to turn the tide in its war with Iraq, that despite US neutrality the Reagan administration would defend its vital interests in the Gulf. Also that year, America created a Rapid Deployment Force which became Central Command (CENTCOM) to deter Russian intervention in the Middle East. By the spring, Iraq had been forced to abandon its offensive completely. Iran recaptured Khorramshahr in May and in June Iraq withdrew the bulk of its forces to the Iraqi border to fight a defensive war. Worryingly, Iran then carried the war into Iraq in July 1982, with an operation launched towards the Iraqi city of Basra.

It has been alleged that Iran timed its offensive against Iraq to coincide with Israel's invasion of Lebanon, thus limiting international censure and frustrating Arab efforts to impose an oil embargo on Israel's allies.

On 6 June 1982, Israel launched Operation Peace for Galilee, invading Lebanon along a 63-mile front. Israel was clearly hoping for a repeat of the Six Day War. It was only designed to force the 15,000-strong PLO out of southern Lebanon, but at the same time it seemed highly unlikely that confrontation with the Syrians deployed in the Bekaa Valley, Beirut, the Shouf Mountains and along the Beirut–Damascus Highway could be avoided. Also, on the central front over Nabatiyeh the first Israeli plane (an A-4) was shot down by a handheld SA-7 and the pilot captured. The Nabatiyeh area was found to contain a major PLO training camp with numbers of international terrorists who were undergoing training (in total, Israel rounded up about 1,800 terrorists in southern Lebanon from twenty-six countries).

The following day, the IAF attacked targets in Beirut and southern Lebanon. One of the Syrians' first losses was a Soviet-supplied MiG-23 shot down over Beirut in a dogfight. Then, when Israeli forces attacked Jezzine, defended by a PLO/Syrian force, there was some controversy as to whether or not the nearby Syrian SAMs fired at supporting Israeli warplanes. A day later, on 8 June the IAF claimed six Syrian MiGs downed. Tension between the two countries quickly escalated.

Inevitably the Israeli advance was soon threatening Syria's position in the Bekaa, for Israel planned to sever the Beirut–Damascus Highway, which would cut the Syrian forces off. Seizing the initiative, a Syrian brigade at Ein Zhalta ambushed the Israeli column assigned this task. The only way to extricate it was with air cover, but this could not be guaranteed because of the nearby Syrian SAMs in the Bekaa. It was now clear the situation would lead to an escalation in the fighting with Syria.

Indeed, Syrian President Assad's forces in the Lebanon were facing a crisis. War had not been officially declared, but those units in Beirut were in danger of being isolated. Syria was suddenly faced with the prospect of losing her foothold in Lebanon altogether. Assad's brother, Rifaat, urged him to send more SAMs to Zahlé in the Bekaa. Defence Minister General Mustafa Tlas opposed this, as his intelligence showed the Israelis could probably knock them out. Regardless, extra batteries were despatched. They also had some 700 pieces of AAA in the valley.

The Israelis saw this as an aggressive move, but since the 1973 Yom Kippur War there was a question over the IAF's ability to cope with a sophisticated in-depth missile defence network. The IAF had learned the hard way in 1973 that, before all else, they must attack radar and SAM sites. Through the use of electronic counter-measures (ECM), the IAF was able to regain dominance in the air in 1973, but not before the Arabs had accounted for 115 Israeli aircraft and many precious pilots. These high losses had brought into question the fighters' ability to operate against dense air defences, although a succession of IAF commanders stated they could not only cope but could destroy them.

Therefore, on Wednesday, 9 June 1982 the elimination of the Syrians SAMs at Zahlé was authorised. Nine years of exercises, meticulous training and great expense were now put to the test. The destruction of the SAMs and the subsequent air battle are a very important episode in the history of aerial warfare. The targets consisted of nineteen batteries of SA-6s deployed at several locations along the Bekaa Valley, which is some 25 miles by 10 miles flanked by ridges up to 6,500ft high. At 2 p.m. the IAF executed a well-co-ordinated attack

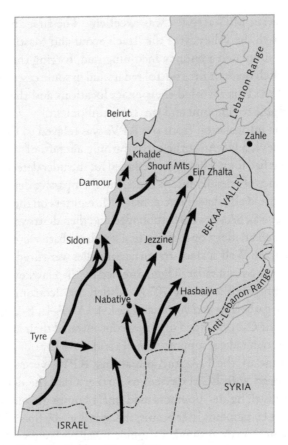

Operation Peace for Galilee, 6 June 1982.

on the Syrians' missile sites with a two-wave attack (comprising ninety-six aircraft followed by another ninety-two).

The Israelis knew every Syrian move because of their Remote Piloted Vehicles (RPVs), which gave their air controllers the ability to observe real-time targets and guide Israeli attack planes more effectively. This gave the Israelis almost total control over the Bekaa target area. Israeli intelligence on Syrian MiG activity was excellent. The first aircraft to approach the valley were the Israeli Scout and Mastiff RPVs. They simulated a phoney incoming raid, forcing the Syrians to switch on their fire control radars and in some cases to actually fire. This revealed their exact locations and the Israelis were able to pinpoint and jam their equipment.

The information from the RPVs was relayed to the E-C2 Hawkeye, a US Airborne early warning aircraft, of which the Israelis had four. They were designed for intruder detection and guidance of friendly interceptors, but their passive defence systems could also detect electromagnetic emitters on the ground. The Syrian SAMs were first blinded and then destroyed.

Israeli artillery opened up first on those batteries in range, and Israeli Wolf surface-to-surface missiles were fired, homing in on the Syrian radar. Then, directed by the Hawkeyes under the control of a Boeing 707 (modified for electronic intelligence gathering and command tasks), IAF F-4s, F-15s, F-16s and Kfir C-2s swooped in. They concentrated their attacks on the Syrian radars, support vehicles and SA-6 sites.

Some of the F-4s and remaining RPVs were specially equipped with ECM devices to further blind the defenders. The Israeli fighter-bombers used smart bombs guided to their targets by pinpoint laser beams, this was done by Special Forces personnel on the ground, TV-guided Walleye bombs, Maverick missiles guided by TV, infrared or radar emissions, Shrike bomb radar homers, and conventional iron bombs. Chaff and flares were also used by the IAF within the Syrian air defence area.

The Syrians manning the missile system were caught completely by surprise; the missile brigade that had just arrived was still deploying. Those who were prepared found if they left their radars on they were destroyed by the various homing ordnance; if they switched them off, they were vulnerable to TV-guided, infrared or laser bombs. Nor were attempts at using the radar intermittently successful, as they could not breach the ECMs. Numerous SA-6s were fired, but none found their targets, though several fighters are reported to have had moderate to severe battle damage.

In the course of this raid, which reportedly took just ten minutes, the IAF destroyed all but two of the SA-6 sites as well as several supporting SA-2s and SA-3s without the loss of a single Israeli aircraft. The remaining two were damaged, while the seventeen destroyed may have included four SA-8 launchers. One SA-8 battery was even knocked out by an RPV carrying an explosive payload. The two surviving SA-6 sites, as well as some additional batteries that were replenished with new equipment overnight, were destroyed by the IAF the following day.

Those missile batteries that escaped destruction were attacked by the second wave of IAF jets, using smart bombs and conventional explosives. There was some danger from the AAA guns deployed in the valley, but they were under Israeli artillery fire and proved useless. The Syrians added to the increasing chaos by releasing smokescreens in a futile bid to hide their positions. In fact, the smoke simply helped the Israelis pinpoint their targets more easily.

Now, in a desperate effort to save their defences, the Syrians committed large numbers of interceptors. The day before they had started to send out big groups of up to twelve planes looking for the smaller Israeli formations. The result was a huge confused air battle involving some 200 jets in an area 2,500km^2. What occurred was not a swirling multiple dogfight in the classic sense, like the Battle of Britain, but a series of independent encounters. At its height, some ninety IAF jets and sixty SAF jets were simultaneously airborne.

The Israelis had their F-15s flying cover for the strike aircraft, which were rapidly vectored in on the incoming MiG

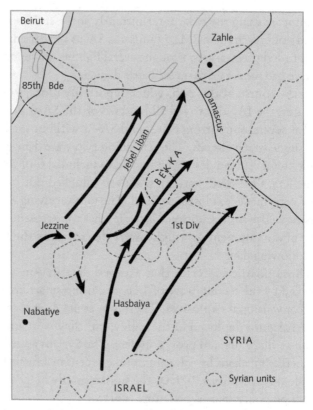

Fighting in the Bekaa Valley, 7–11 June 1982.

and Su fighter-bombers by the ever-watchful Hawkeyes. Although the F-15s were exclusively assigned to air-to-air cover, they were soon joined by other IAF aircraft, in particular the versatile F-16 ground-attack and air-superiority fighter.

In sharp contrast to the Israelis' flexibility, the Syrians adherence to Soviet training meant they could only operate efficiently if they stuck to the manual. They were used to having ground control and SAM support in an integrated air defence system. None of this now existed over the Bekaa due to the SAMs' destruction and the ECMs. What followed was a Syrian disaster. As the fighters became mixed up at incredible speeds, the Syrian AAA gunners had to cease fire and both sides' pilots were faced with the problem of target identification. For the Israelis, it was to be a turkey shoot.

The F-15s achieved most of the IAF's air-to-air victories over the Bekaa, although one F-16 pilot may have downed as many as five Syrian MiGs. Most of the Israeli kills were due to the Sidewinder AIM-9L air-to-air missile (AAM) and the Israeli modifications, the Shafir 2 and Python 3 AAMs. In terms of weaponry and guidance systems, the Israelis were better equipped. The Syrians' Soviet systems had smaller lock-on envelopes, and even their newer Advanced Atoll and Aphid suffered from this. There were even reports of Israeli gun kills using just the aircraft's cannon; this was some achievement, and clearly illustrated Israeli pilots' superiority.

By the end of the first encounter with the SAF, the Israelis had shot down twenty-nine MiGs. Overall, 30 per cent of the Syrian warplanes that flew into Lebanese airspace were damaged or destroyed. In three main air battles fought during the day, the Israelis would claim forty-one Syrian aircraft shot down for the loss of no Israeli fighters.

On Thursday, 10 June, another air battle cost the Syrians a further fifteen to twenty-five MiGs and four helicopters. The next day, the Israelis and Syrians agreed to a ceasefire, but before it went into effect a final air battle was fought in which the SAF lost another eighteen MiGs.

In total, the SAF lost ninety-one aircraft, including MiG-23s/25s and six helicopters, in three days of aerial fighting. Syrian losses would have been even greater had Israeli pilots been allowed to pursue their opponents over Syrian airspace. According to United States Air Force figures, about forty Syrian fighters were downed by F-15s, forty-four by F-16s and one by an F-4. The AIM-9L air-to-air missile accomplished most of the kills. Also, on the ground, Israeli AH-1S Cobra and Hughes Defender helicopters knocked out twenty-six tanks, sixteen armoured personnel carriers (APCs) and thirteen other vehicles during the invasion.

Not a single Israeli warplane was shot down in air-to-air combat. They only lost three jets – an A-4, F-4 (S) and a Kfir – all of which were brought down by ground fire. A Russian politician subsequently claimed the IAF had lost up to forty aircraft. At least four Israeli helicopters were lost to Syrian ground fire and two to friendly fire. Syrian Gazelles claimed two tanks and one M-113 APC using missiles.

It is unlikely the Syrian General Staff and Air Force Command knew what was actually happening over the battlefield and did not realise the size of the disaster until Wednesday evening. The defeat must have been a particular blow to President Assad, who was a former air force general. The SAF lost about 25 per cent of its total force and many of its aircrew. The real losses, though, were the precious pilots; aircraft can be easily and relatively quickly replaced. About 50 per cent of the pilots shot down were either killed or put out of action for good, meaning the SAF lost forty to fifty combat pilots. After the battle over the Bekaa, the SAF made only occasional appearances and their ground forces were left without air support for the rest of the war. However, Syrian Gazelles were used to some effect.

On the ground, although many of the Syrian Amy's units had been severely battered in the often fierce engagements with the IDF, the bulk of them in the Bekaa were able to withdraw to the vicinity of the Damascus Highway. A cease-fire with the Syrians, including the PLO, went into effect on 12 June. Sporadic fighting continued in the Bekaa and on

22 June SAM sites were again attacked and two SAF MiG-23s shot down. A month later on 24 July, after it had been discovered that three batteries of SA-8 had been brought in the IAF destroyed them. Also, an Israeli F-4 flying a reconnaissance mission was shot down by an SA-6.

The Soviet Union began resupplying Syria with replacements only a day after the Bekaa air campaign commenced. As well as replacing the lost MiGs and SAMs, they provided several new weapon systems including SA-8s, SA-9s and SA-5s. An IAF F-4 was downed by one of the SA-8s in the Bekaa in August 1982, during a raid designed to eliminate the new systems. A subsequent attack was launched by another F-4 to destroy the wreckage and prevent any ECM equipment falling into Soviet hands; Soviet technicians were allegedly killed during the raid. By the end of September, the IAF had destroyed twenty-nine SAM sites in seven raids with only two IAF aircraft lost to ground fire.

The PLO received a knockout blow. At Damour the IDF captured 5,000 tons of arms and ammunition, including SA-9 missiles. By the end of June, the IDF had collected 4,170 tons of ammunition, 764 vehicles, 26,900 light weapons and 424 heavy weapons. Somewhat ironically, in August Israel sold $50 million worth of captured Syrian and PLO arms and ammunition to Iran. To some, this confirmed their collusion.

The IDF besieged the PLO in Beirut until early August when they finally agreed to quit the city; some 14,398 Palestinian fighters and Syrian soldiers withdrew. Syrian Army losses are believed to have been up to 1,200 dead, approximately 3,000 wounded, 296 prisoners and almost 500 armoured vehicles. The PLO lost 1,500 dead, an unknown number of wounded and, more importantly, their entire infrastructure (including their archives) in Lebanon. Israeli casualties, however, were high. Operation Peace for Galilee cost the IDF 368 dead and 2,383 wounded in six weeks of fighting.

Although Israel won the battle and achieved her immediate objectives with the expulsion of the PLO from Lebanon and neutralisation of the SAF, it was much harder to win a war of occupation. Within two years, Israeli casualties had almost doubled and by 1984 they had suffered 600 dead, proportionally equivalent to half the total US death toll in Vietnam. The number of wounded rose to 3,600, again proportionally much higher than those suffered by the US in Vietnam.

The indigenous Shia Muslim Lebanese slowly turned on the IDF once it was apparent that it was not going to withdraw after ousting the Palestinians. Israel wanted to put in place safeguards to ensure that the PLO could not return. This saddled the Israelis with a security zone in southern Lebanon, which was to be a political and military headache for almost twenty years. The IDF finally withdrew from southern Lebanon in May 2000.

The Soviet Union was naturally puzzled by how easily the Syrians' Soviet-supplied air defence system was destroyed and no doubt rather embarrassed. Israel has kept secret its true methods, which means that accounts of the ease with which the system was overwhelmed have to be viewed with some scepticism. It was not the triumph of a single technological system or weapon, but a combination and its expert employment. Therefore, only a partial picture of the method of attack can ever be drawn. Nonetheless, the air battle over the Bekaa remains a dramatic and important engagement in the development of modern air warfare. It also remains widely acclaimed as the largest single air battle since the Second World War. Certainly, the confrontation was the largest in the history of Middle Eastern warfare.

19

A Show of Force: Khafji 1991

The Battle of Khafji and, in fact, the whole of the Desert Storm campaign was to exemplify the decisive role of air power on the modern battlefield. The campaign is also viewed as the first truly electronic war because of the technological sophistication used to prosecute the conflict.

Although it was anticipated that the Gulf War would be undertaken largely in the deserts of Saudi Arabia and Iraq, there was concern that a major urban battle would have to be fought to liberate Kuwait City. A taste of potential things to come occurred on 30 January 1991 when an unexpected urban engagement was fought in the Saudi town of Khafji.

Spearheaded by two Republican Guard Corps armoured divisions, Iraq invaded Kuwait on 2 August 1990. They advanced on Kuwait City, as did a Special Forces commando division. Heliborne Iraqi units also seized strategic sites, including the island of Bubiyan. The following day, a mechanised Republican Guard division moved to secure Kuwait's border with Saudi Arabia, sealing the country off from the outside world and help. Approximately 140,000 Iraqi soldiers and 1,800 tanks poured into Kuwait. The Kuwaiti Army, not fully mobilised to its standing strength of just 16,000, was swiftly overwhelmed. Kuwait found itself under Iraqi control within just twelve hours. Only around the emir's palace in Kuwait City was there any extensive resistance lasting about two hours.

In the wake of the invasion, Western and Arab states were quick to deploy forces to defend Saudi Arabia. Ever since the loss of the pro-Western Shah of Iran in 1979, Saudi has been viewed as the major Western ally in the region. The Gulf crisis and Operations Desert Shield/Storm pushed the Saudi Arabian Armed Forces (SAAF) to the fore. The Saudis face a permanent demographic problem: they have insufficient manpower to deter their more populous neighbours, Iran and Iraq. In fact, to counter potential threats Riyadh embarked on one of the most expensive procurement programmes outside the superpowers. The Saudis increased defence spending every year after the outbreak of the Iran–Iraq War. Over the last two decades of the twentieth century, Saudi Arabia conducted a steady military expansion to enhance her security. However, the end of the Iran–Iraq War 1980–88 did not herald a halt to the regional arms build-up. Saudi defence expenditure stood at about 30 per cent of GDP, dipping in the mid-1980s and mid-1990s due to the fall in oil prices.

The US has remained Riyadh's largest arms supplier, but the pro-Israeli lobby within the US Congress strained this relationship. While Saudi Arabian military support for the Arab–Israeli Wars was small, 4,500 men in 1967 and 1,500 in 1973, financial support was considerable, as was support for the PLO. Although there was no formal agreement, it was accepted that the 300,000-strong US CENTCOM – directed from Florida – or the Rapid Deployment Force, could use Saudi bases to defend the Gulf in the event of a crisis. This

proved more than prudent. In the event, it took US intervention, as expected, to guarantee Saudi sovereignty in the face
of Iraqi aggression against Kuwait.

In response to the Kuwaiti crisis, the US instigated
Operation Desert Shield, a massive multinational exercise
to defend Saudi Arabia in the event of Iraq moving further
south. The Coalition gathered half a million men from thirty-
one countries, comprising nine US divisions, one British, four
Arab and one French. They were equipped with 3,400 tanks
and 1,600 pieces of artillery. The Allied air forces included
1,736 combat aircraft and 750 support aircraft. America's
commitment to this and the subsequent Operations Desert
Storm/Sabre to liberate Kuwait was staggering, consisting of
280,000 troops with 600 tanks, 80,000 navy, 50,000 air force
and 90,000 marine corps with 200 tanks. Britain and France's
commitment was not inconsiderable, with 33,000 men and
120 MBTs and 11,500 men with forty MBTs respectively.

The Arab Gulf Co-operation Council only deployed
10,000 men, although in the Gulf region they had available
150,000 men and 800 tanks. The largest Arab contingents
came from Egypt, with 47,000 troops and 500 tanks and Syria
with 19,000 men (in return for its support, Syria received
$1 billion from Saudi Arabia). Both countries stated their
troops were only deployed to defend Saudi. Nevertheless, with
the commencement of the ground war to liberate Kuwait on
24 February 1991, Egyptian and Syrian forces were committed
to the offensive on the proviso they were not used inside Iraq.

Other forces included the Kuwaiti Army in exile – three
or four brigades, totalling 10,000–15,000 men, equipped
with Chieftain and Yugoslav M-84 MBTs (Soviet T-72s
made under licence). About 17,000 members of the Kuwaiti
armed forces were allegedly held in Iraq. From Bangladesh
there were 6,000 troops, 5,000 from Pakistan, 1,500 from

Morocco, 480 from Niger and 500 from Senegal. Saudi King Fahd was commander-in-chief SAAF, although operational control of all Arab forces came under his nephew, Lieutenant General Khalid bin Sultan, son of Prince Sultan ibn Abdul Aziz, the Saudi defence minister.

After the failure of Saddam Hussein's wars against the indigenous Kurdish separatists in northern Iraq and the futile eight-year war with Iran, he had sought to distract internal politicking by seizing Kuwait's oil reserves. Iraqi forces deployed in the Kuwaiti Theatre of Operations (KTO) at the beginning of January 1991 allegedly numbered 540,000 men, armed with 4,000 tanks, 2,700 APCs and other armoured fighting vehicles (AFVs) and 3,000 pieces of artillery. This force included the 120,000-strong Republican Guard Corps stationed to the north, straddling the Iraq–Kuwait border.

Saudi had two main forces, the regular armed services and the National Guard, whose role is to protect the House of Saud. By 1991 the SAAF numbered 67,500. The army, or Royal Saudi Land Forces (RSLF), fielded 40,000, organised into two armoured, four mechanised, one infantry and one airborne brigade (including 10th Armoured, 8th, 10th, 11th and 20th Mechanised Brigades), equipped with 550 tanks and 1,840 combat vehicles, APCs and armoured cars. Paramilitary forces had been expanded because of the threat of fundamentalist subversion and Iranian inspired riots. The Saudi Arabia National Guard (SANG) totalled 55,000: 35,000 active and 20,000 tribal levies, equipped with 1,100 American V-150 Commando APCs.

The Royal Saudi Air Force (RSAF) numbered 18,000, equipped with 189 combat aircraft, including forty-four British Tornadoes, forty-two American F-15s and fifty-three F-5s. Rather embarrassingly, a pilot defected with his F-15 fighter to Sudan in November 1990. Sudan kept the aircraft for several days, during which time Iraqi intelligence would have had a chance to examine it closely, gaining frequency information about electronic warning, electronic counter-measures and communications. This confirmed the US Congressional pro-Israeli lobby's fears over selling Saudi the equipment in the first place. Even so, by the end of January 1991 the RSAF had flown 2,000 sorties in support of Desert Storm and by 28 February had lost three aircraft (two F-5E Tigers and one Tornado IDS).

The Coalition's combined air forces launched a round-the-clock air campaign, targeting Iraqi command and control sites, Scud missile installations, lines of communication, the Republican Guard MBTs and APCs. Vitally, air superiority over Kuwait and Iraq was achieved within just twenty-four hours of the first air attacks. Some 35,000 air sorties were launched against Iraqi ground forces, of which 5,600 were directed at the Republican Guard. The Americans also committed sixty-two B-52 bombers. These veterans of the Vietnam War flew 1,600 sorties and accounted for 30 per cent of the tonnage dropped. Initial estimates of Iraqi losses to air strikes were about forty Republican Guard tanks (the guard were dug in over a 4,000-square-mile area) while the army lost fifty-two tanks, fifty-five artillery pieces and 178 trucks. According to CENTCOM, Iraqi execution battalions were patrolling front-line troops shooting deserters or anyone listening to Coalition media.

The Saudi town of Khafji (also al-Khafji, Al-Khafji or Ras al-Khafji), about 12 miles south of the Kuwaiti border, first came to public notice on 16 January 1991 when an oil storage tank was struck by Iraqi artillery. Khafji is the only junction on the coastal road linking Saudi Arabia with Kuwait, to the north, and Bahrain, Qatar, the UAE and Oman to the south. This attack heralded Saddam's campaign of environmental warfare during which he was to ravage Kuwait's oilfields. The shelling caused a fire and a cloud of black smoke lingered over the town like some ominous warning. In light of the danger, the population of 15,000 were quickly evacuated. The whole world knew that Khafji had been empty for two weeks after global news networks broadcast scenes of the dust-blown streets.

Saddam clearly decided to pre-empt the ground war, possibly in the hope of provoking the Coalition before it was ready. If he seized the initiative and generated enough momentum, he could suck in Coalition forces piecemeal, inflict casualties and secure prisoners, thereby damaging Coalition morale and unity in the glare of the global media. In particular, he turned his eye on the US Marines.

Khafji was well within the range of Iraqi artillery, sited beyond the Kuwaiti border, as had been proved by the shelling of the oil installation. This meant that Coalition forces were obliged to remain far to the south of the town, in effect, leaving it undefended. It was a vacuum that Saddam intended to fill. Luckily, as Khafji had been evacuated there was no risk of civilian casualties, the town was open plan with mainly two-storey buildings. Therefore, rather than providing a defenders' paradise, it actually gave them nowhere to hide.

It seemed a foolhardy operation, but at the time it presented a good opportunity to Iraqi planners. The town was close to Iraqi defensive forces within the KTO that could provide artillery cover; it was within easy reach of the Iraqi Air Force; and being on the coast the Iraqi Navy could also provide support. Furthermore, the area was defended largely by the Saudi forces, which the Iraqis may have felt would be a soft option. The neighbouring US Marines were to be drawn into the town and pinned down by a frontal attack. A flanking tank force was then to strike them from the west, while to the east troops were to be landed by the Iraqi Navy. The Iraqis were to make, in total, four forays along the border from Wafra eastward to Khafji on the coast.

Across on the Kuwaiti side of the border, Iraqi forces consisted of the 5th Mechanised Division. Its parent corps planned to launch a simultaneous three or four-pronged assault. It is unclear if it was part of a wider attack down the Wadi al-Batin, although the lack of adequate air cover made this unlikely. To reinforce the attack on Khafji, three other mechanised divisions, some 60,000 men and 240 tanks, were gathered near al-Wafra in Kuwait. It is notable that the elite Republican Guard was not assigned any part in this operation. The Iraqi Air Force, the other key player in the affair, was also to be conspicuous by its complete absence.

Six days after the bombardment of Khafji, an Iraqi armoured division was detected deploying a convoy towards Kuwait. Organisationally, Iraqi mechanised divisions had two mechanised and one armoured brigade; armoured divisions had two armoured and a mechanised brigade; while infantry divisions had three infantry brigades and a tank battalion. Armoured brigades had three armoured and a mechanised battalion, while mechanised brigades had three mechanised and a tank battalion. Divisions usually have four artillery battalions in support. The Coalition response was swift and an air strike claimed fifty-six of the vehicles. Normally, once a hit was scored the Iraqi crews of any neighbouring vehicles would abandon their equipment and run, leaving them to be picked off.

Elements of the Iraqi Army's 5th Mechanised Division, up to brigade strength of three to four battalions, some 2,000 men, with forty to fifty T-54/55 main battle tanks and BMP/YW 531 tracked APCs, moved out over a 50-mile front on the night of 29 January. They crossed the Kuwaiti border undetected and pushed south, supported by fast patrol boats, which moved down the coast bearing landing parties.

The RAF pounced on the naval element of the assault, hitting at least two of the vessels and scattering the others. The Royal Navy also attacked some Iraqi patrol boats, possibly intended for the Khafji operation on 29 January. A British frigate detected the boats and launched Lynx helicopters armed with Sea Skua missiles. A second convoy appeared the following day, including an Iraqi minesweeper, three ex-Kuwaiti fast-attack boats and three landing ships. RAF Jaguar and USAF A-6 jets attacked them. In the desperate Battle of Bubiyan Island, twenty Iraqi naval craft attempted to flee to Iranian waters. Chase was given and only two damaged vessels, a patrol boat and an amphibious ship managed to escape the carnage. During the whole war, the Iraqi Navy was to suffer 143 out of 165 surface combatants damaged or destroyed.

In the face of Saddam's incursion, the Coalition's first move was to cut his lines of communication between Kuwait and Khafji. British Jaguar jets and American A-10 Thunderbolt tank-buster aircraft were despatched to attack any Iraqi forces north of the town. American Marine Corps Cobra attack helicopters and artillery were also used to seal the town off. About 55 miles from the coast, a supporting Iraqi tank brigade (possibly the 26th) ran into the 1st US Marine Division's light armoured infantry battalion. In the firefight that followed, involving American attack helicopters and A-10 tank busters, the Iraqis lost twenty-four tanks and thirteen other vehicles. An Iraqi tank managed to account for one marine light armoured vehicle, while a second was destroyed by friendly fire. The Americans suffered eleven fatalities, of whom seven were by friendly fire. This resulted in aerial recognition signs, in the shape of an inverted white 'V', being painted on the side and an orange marker placed on top of every Coalition vehicle.

A second supporting Iraqi brigade, possibly the 20th Mechanised, bounced off the 2nd US Marine Division's 2nd Light Armoured Infantry Battalion. One LAV with tube-launched, optically tracked, wire-guided (TOW) missiles accounted for two Iraqi tanks. To the west of the town, the Iraqi 15th Mechanised Regiment ran into a Qatari tank unit. Iraqi forces also suffered an air attack and fell back, losing eighty vehicles outside Khafji. Remarkably, the Iraqis' remaining armoured battalion was to push aside a screening Saudi force and occupy Khafji, which was only lightly defended by Saudi and Qatari forces.

Late on 29 or early 30 January, the Saudis heard that fifty-seven Iraqi armoured vehicles were heading towards Khafji's desalination plant. A company of lightly armed Saudi marines was quickly withdrawn from the town. The initial slowness of Coalition airpower to respond to this threat was due to the American marines struggling to turn back elements of the 20th Mechanised Brigade and 26th Armoured Brigade further west. During the early hours, a flight of Cobras from 369th Gunfighter Wing hunted the Iraqi armour with night-vision goggles. They ran low on fuel and had to return to base and were replaced by four more Cobras, which destroyed a platoon-sized mechanised force. By 5–6 a.m., the Saudis realised the Iraqis really meant business, as about twelve Iraqi tanks or APCs were observed on the western edge of the town. Nevertheless, no Iraqi tanks entered Khafji; they were all knocked out to the north, although some armoured vehicles did get in.

The following day, under attack from A-10s, the other two attacking Iraqi battalions failed to get through. Inside Khafji, although a brigade of 2,000 had been thrown into the attack, the Iraqis only had about 600 defenders with infantry support weapons and some tanks on the outskirts. This solitary battalion was to bravely hold out for two days. Although the town had no real military value, the Coalition could not leave them, there would have been a loss of face, plus a US Marine reconnaissance party was trapped inside Khafji along with the Iraqi soldiers.

If the Coalition had given the Iraqis a breathing space there is every reason to believe that they could have turned Khafji into a stronghold defended by minefields and interlocking strongpoints, incorporating the town's buildings. The Coalition knew from the Iran–Iraq War that the Iraqi armed forces were moderately competent in the offensive; of greater concern was the fact that they had fought eight years of almost continuous defensive action against the Iranians. The Iraqi Army was skilled in constructing defensive positions, berms, minefields and laying down artillery fire. Furthermore, the Iraqi Navy might attempt to replenish the town under the cover of darkness.

The Battle of Khafji was to be one of national prestige. Because this was Saudi soil, it was felt that the Saudis with support from Qatari troops should be given the role of driving the Iraqis out. The assault force included three mechanised SANG battalions and elements of a Saudi armoured battalion with supporting American M60A3 tanks, backed by a Qatari armoured battalion equipped with about twenty French-supplied AMX-30 tanks. The Qatari tanks were assigned seven tanks to each company of the 7th Battalion, 2nd SANG Brigade. The Qataris were to be kept in reserve, their lack of communication with the Saudis, due to radio incompatibility, meant it was difficult to co-ordinate their efforts. These forces were under Saudi General Khalid, backed by US Marines and airpower.

The use of the SANG may have been due to it being the most politically powerful arm of the Saudi military, or because of its strong tribal roots, or the fact that the situation was arguably an internal security problem. Also, it was perhaps because it was considered more efficient than most Saudi Army formations. However, lacking tanks the SANG could not have acted on its own.

Any counter-attack trundling across the flat open ground would be instantly spotted by Iraqi lookouts on the perimeter of the town. Indeed, on their approach the Saudis and Qataris found themselves under quite intense artillery shelling. American counter-battery fire detected and silenced the Iraqi guns. Ironically, two American marines recon teams, some twelve men, who remained trapped in the town, acted as artillery observers for the besieging troops. They had at least one near miss when an Iraqi patrol took shelter in a

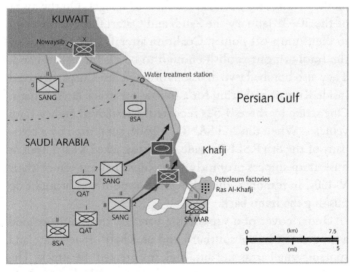

Khafji, 30 June 1991.

neighbouring room for several hours. Fortunately for the Coalition, the marines were not discovered and continued to direct friendly fire into the environs and heart of Khafji.

The Saudis were slow and hesitant, although they blamed the lack of air support by the US Marines. Once the Cobra, Harrier and A-10 strikes got going they accounted for thirty to thirty-five tanks, alleviating the threat to the relief force. These air attacks also served to slow down the Saudis, who withdrew 2–3km every time there was an air engagement. On their initial advance, the Saudis successfully secured the surrender of five Iraqi tanks and six APCs outside the town, their crews probably demoralised by the airstrikes. Other units were not so keen to give up and one Saudi V-150, equipped with TOW, was credited with knocking out eight Iraqi tanks.

It was decided to launch the first assault under the cover of darkness, potentially a tricky operation despite the use of modern technology such as the thermal imager. On the night of the 30–31 January the Saudi and Qatari forces moved up to their jump-off points. Coalition aircraft silenced most of the Iraqi armour foolish enough to be visible beforehand. They also bombed two divisions of Iraq's 3rd Corps, detected inside Kuwait gathering for a follow-on attack at Al-Khafji. One strike by three B-52s reportedly destroyed eighty Iraqi vehicles. When the 2nd SANG Brigade, supported by a company of the 8th RSLF Brigade, finally attacked Khafji itself at dusk Iraqi snipers promptly immobilised about ten of their V-150s. In response, the 8th RSLF's armour was brought up pushing the Iraqis back.

Under cover of a supporting barrage, the Saudis gained a lodgement in the southern end of Khafji. Inaccurate and sporadic small arms and anti-tank fire greeted the attack. The Iraqis, recovering from the first clash, proved to be resilient and some of their remaining soldiers had to be cleared from the town's outlying buildings one by one. Nevertheless, in the chaos, others quickly threw down their arms, perhaps realising they were cut off and that resistance was now futile.

The main counter-attack was launched the following day with two prongs: the 8th Mechanised Infantry Battalion/ 2nd SANG, reinforced by the Qataris tanks and HOT Anti-Tank Platoon on the right, while the 7th Battalion/2nd SANG was on the left. Saudi marines also moved up the coast road to secure the south. In the confusion, the Qataris opened fire on the SANG, although luckily there were no casualties.

On the second day of the Battle of Khafji, the Saudis discovered their M60 tanks outgunned the T-55s by 900m and their TOWs by 2km. The 7th Battalion, 2nd SANG, advanced on the south of Khafji supported by the 3rd Marines and the Qataris. In the first clash inside Khafji itself the Saudis suffered one killed and four wounded. The Iraqis managed to knock out three Saudi armoured vehicles with RPG-7 anti-tank weapons. The 7th and 8th SANG Battalions forced their way further into the town, fighting street to street for the rest of the morning. Iraqi forward observers operated from Khafji's water tower and as a result took a lot of incoming fire. Five Iraqi dead were found in and around it. During the fighting, the Iraqis had called repeatedly for air support but the Iraqi Air Force stayed away. Coalition forces lost nineteen dead and thirty-six wounded.

Saddam's follow-on forces at Wafra, as well as a 10-mile-long column passing through Kuwait, were attacked by Coalition aircraft, suffering heavy losses, and were forced to retire. The American 1st Armoured Division also used the Advanced Tactical Missile System on Iraqi targets. At Wafra, the Iraqis lost over seventy tanks and APCs, while the Americans captured eleven Iraqi POWs and suffered eleven dead and two wounded. Had Saddam Hussein wanted to bring up his Republican Guard as reinforcements, they were under aerial siege. In a twenty-four-hour period they endured 350 tactical and ten B-52 strikes. On 1 February, they were on the receiving end of another 600 sorties.

To the north of Khafji, an Iraqi brigade with 100 armoured vehicles tried pushing south. The Saudi TOW Platoon from the 8th RSLF Brigade engaged them for thirty minutes, destroying twelve of their vehicles. They were then harried back by Coalition air attack, and those Iraqis fleeing north from Khafji lost another dozen tanks to airstrikes. The US Marines countered an Iraqi armoured battalion making an attack 50 miles to the west, near Umm Hujul.

The third day of fighting for Khafji lasted twelve hours. When eventually the Iraqi snipers had been silenced, it was discovered that the Iraqis had lost in and around the town at least forty dead (some sources record as many as 300), thirty-five wounded, 463 prisoners and forty to eighty armoured vehicles. Several of the Chinese-built YW 531s were found knocked out in the streets, lost either to the Coalition's fighter-bombers or artillery. The whole Iraqi garrison had been completely destroyed.

In the three days, the Saudi forces lost eighteen men and thirty-two wounded, three tanks and a TOW launcher. During that period, according to the Saudis, the Iraqis lost eleven tanks and fifty-one APCs destroyed, and nineteen APCs captured. The Saudis bagged a total of 1,040 POWs.

First blood had gone to the Coalition forces and the SANG had given a good, if patchy, account of itself. Furthermore, the Coalition was able to heave a sigh of relief that the urban battlefield had not turned into a costly bloodbath and a public relations disaster. This proved a major morale boost for the untried SAAF and showed they could conduct themselves well. A Saudi spokesman, Colonel Ahmed al-Robayan, stated the success of the combined arms operation against the Iraqis at Al-Khafji was a 'very good sign', as it tested command and control of the multinational force. 'The Saudi forces are doing an outstanding job,' said Marine Brigadier General Richard Neal, deputy director of CENTCOM operations. 'They did a terrific job at Khafji ... I've been impressed with them.'

Jubilant Saudi forces were in the fore when the ground phase of Desert Storm started in the early hours of 24 February 1991. Just before the ground offensive, the Coalition was claiming the destruction of up to 1,300 tanks, 800 armoured personnel carriers and 1,100 artillery pieces. The Iraqi Air Force lost almost 50 per cent of its holdings, some 272 aircraft: 141 were claimed destroyed, while the rest fled to Iran never to return. Iraqi forces in the KTO were thought to number about 250,000, of which about 150,000 were actually inside Kuwait. It is estimated the Iraqis numbered about 200,000 when the ground war was launched, the air war was initially thought to have accounted for 100,000!

In an admirable example of international co-operation, eleven nations took part in the offensive to liberate Kuwait. Joint Force Command East (comprising Saudi, Kuwaiti, Omani and UAE forces) pushed into Kuwait to form the anvil for the American, British and French thrust into Iraq, which effectively trapped the bulk of the Iraqi forces in the KTO. The SAAF and Kuwaiti forces were almost in Kuwait City by 26 February, when the Iraqi garrison abandoned it, heralding the beginning of the end for the Iraqi Army in Kuwait. They agreed to a ceasefire two days later. Kuwait estimated that 1,000 civilians were murdered during Iraq's occupation.

When the fighting was over the air war death toll was revised to 10,000. The Iraqis themselves claimed they had lost 20,000 dead and 60,000 wounded in twenty-six days of air attack. It was also estimated that another 10,000 were killed during the land offensive. Despite this, a total of 20,000–30,000 dead seems rather high and a conservative estimate of below 10,000 is more likely. Approximately 85,000 prisoners of war were taken and total Iraqi tank losses were up to 4,000.

Although only a minor engagement in the overall scheme of things, Khafji echoes German attempts to move their armour to the Normandy bridgehead in 1944. If the Iraqis could have widened the battle by bringing up its follow-on forces and offered some token air cover, the fight may not have gone so easily for the Coalition. It is possible the Americans would have had to divert more troops to help the SAAF, distracting them from the left hook. Without air support, the Iraqi offensive was totally compromised and from that point on they fought a largely defensive battle. It was air power and artillery that were to dominate at Khafji and were a dominant feature of the whole campaign.

The Battle of Khafji did prove that the Iraqis could achieve limited tactical surprise and their stout resistance once in the town was a propaganda victory for Saddam Hussein. It took the Saudis three days to fully secure Khafji and they quickly discovered their wheeled armoured vehicles were inadequate for the job. Nonetheless, Saddam's show of force failed with heavy losses and proved that the Iraqi Army was extremely vulnerable to air attack. His gamble did not pay off. If Saddam had picked a better place, concentrated his forces quicker and attacked more effectively, he might have seriously disrupted the Coalition's plans. However, the final outcome was never really in any doubt.

20

ASYMMETRIC WARFARE: TORA BORA 2001

The siege of Tora Bora was the last major battle fought in the opening round of the counter-terrorist war in Afghanistan. This unusual conflict saw American and British air and Special Forces assets directed at unseating the pro-terrorist Taliban Government and destroying the Al-Qaeda terrorist organisation. It was to take two weeks of guerrilla fighting and relentless American bombing to clear the caves and tunnels of Tora Bora, finally ending Al-Qaeda's foothold in Afghanistan.

The world watched in horror when terrorists crashed two airliners into New York's World Trade Center, one into the Pentagon and a fourth en route to who-knows-where on 11 September 2001. These attacks were a major change in the late twentieth-century terrorist's modus operandi. More traditional airline hostage-taking for political reasons saw the aircraft itself become an effective weapon in asymmetrical warfare. Islamic terrorists signalled that no longer was diverting an aircraft and/or blowing it up on the tarmac sufficient to further their ends. The airliner had become a force multiplier. In addition to the terrorists, the finger of blame was also firmly pointed at international security standards and the airlines themselves. The very next day, United Nations Security Council Resolution (UNSCR) 1368 and General Assembly Resolution 56/1 called for international co-operation to bring to justice the perpetrators. They also called for much broader co-operation against terrorism.

The UN had been demanding that Afghanistan stop giving sanctuary to Saudi Osama bin Laden, his Al-Qaeda terrorist organisation and other terrorist groups since the end of 2000. However, 9/11 was a harsh reminder of the failure of the world to act effectively in concert against the growing threat posed by extensive international terrorist networks. An attack of this magnitude and visibility against air travel could simply not be ignored.

The UN had already acted against Afghanistan, or more precisely the Taliban, for harbouring Osama bin Laden and those implicated in the attacks on the Khobar Towers in Saudi Arabia, the US embassies in Africa, the USS *Cole* off Yemen and the earlier attempt to blow up the World Trade Center during the 1990s. UNSCR 1267, which came into effect in November 1999, banned the Afghan carrier Ariana Airlines from operating overseas and required members to freeze Taliban overseas assets. The second, UNSCR 1333, passed by the fifteen-member council with only two abstentions (China and Malaysia) came into effect on 19 January 2001. This followed a month's grace to allow the Taliban to comply with demands to hand over Osama bin Laden and close down camps being used as terrorist training centres. The Taliban's acquiescence of the presence of Al-Qaeda and its complicity in the country's drug trade had ensured that Afghanistan was an undesirable state.

In pursuit of punishing those responsible for the attacks on the World Trade Center and the Pentagon, America decided to form a coalition and use air power and Special Forces to assist the Northern Alliance to oust the Taliban and destroy Al-Qaeda's presence once and for all. The end game for the air campaign was the complete rout of the Taliban regime and the creation of a broad-based successor government under UN auspices to meet the needs of all Afghanistan's ethnic groups. Otherwise drugs, war and terrorism would remain a way of life for another twenty years. In light of their victories over the Northern Alliance and previous experience against the Soviets, the Taliban were not expected to crumble easily.

Operation Enduring Freedom was controlled by US CENTCOM at McDill Air Force Base, Florida. US Naval (USN) forces deployed to the region included F-14 Tomcat and F/A-18E/F Super Hornet fighter-bombers, while global assets included B-1, B-2 and B-52 strategic bombers. The US was not able to conduct any regional land-based fighter missions because crucially Saudi Arabia and Pakistan refused to allow any attacks to be conducted from their soil. Pakistan's government, with its Pashtun population who were linked to the Pashtun Taliban, had to walk a diplomatic tightrope.

As punishment for continuing to harbour Osama bin Laden and his Al-Qaeda terrorist network, air strikes using USN and strategic air assets as well as ship and submarine-launched Tomahawk Land Attack Missiles (TLAMs) were to hit terrorist-related facilities, the Taliban's military infrastructure and their field forces with impunity. Notably, the ground-hugging cruise missile had originally been designed during the Cold War to deliver a tactical nuclear payload by flying under the enemy's radar. It formed part of a graduated escalation to full nuclear war involving intercontinental ballistic missiles. However, they evolved into such weapons as the Tomahawk subsonic cruise missile, designed to deliver a conventional payload. It and the Hellfire air-to-ground missile (later fired from the Predator and Reaper unmanned aerial vehicles) became key weapons in the War on Terror.

The Afghan economy was in ruins well before the air campaign started. The continuing civil war during the late 1990s prevented any recovery. Under the Taliban taxation ceased, foreign investment was almost non-existent and there was rampant inflation. The situation was exacerbated in 1998 when Saudi Arabia withdrew essential financial support from the Taliban for harbouring Osama bin Laden. By then he was wanted for various attacks on US interests in Kenya, Saudi Arabia, Tanzania and Yemen.

The American-led air war opened on 7 October 2001, with bombers flying from Diego Garcia and from Whitman Air Force Base in Missouri. The latter flew on to the island of Diego Garcia in the Indian Ocean after a record forty-four hours in the air and six air-to-air refuels. These were joined by carrier-based fighter-bombers operating from the USS *Carl Vinson* and *Enterprise*, some fifty Tomahawks were also launched.

Since its foundation, the Taliban Air Force (TAF) had been quite active, flying over 150 sorties during the campaigns to capture Taloqan, the Northern Alliance's capital. However, with the loss of Bagram Air Base in 1998 the TAF was forced to destroy or disable many of its aircraft during its retreat from the Shomali Plain. Against the Coalition air campaign, the TAF initially had about eight MiG fighters, eight Su-22 ground-attack aircraft, some helicopters and about four light trainer jets. Most were destroyed on the ground in the opening hours of the air attacks.

At least 50 per cent of the targets hit were terrorist related. These included training camps, stores, safe houses and mountain hideouts. The camps were used to train Chechens, Kashmiris, Pakistanis, Saudis, Tajiks, Uzbeks, Uighurs and Yemenis. Their services were then exported back to their home countries. The bulk of the terrorist facilities attacked were in the Kabul, Kandahar and Jalalabad areas and belonged to Al-Qaeda and the Islamic Movement of Uzbekistan. In particular, Kandahar and Jalalabad remained the favoured targets, as this was where Osama bin Laden and his cronies were most likely to have gone to ground. Kandahar was the stronghold of bin Laden's one-time ally, Taliban leader, Mullah Mohammed Omar.

Initially the air campaign did not greatly affect the Taliban's rudimentary infrastructure or ability to wage war against the Northern Alliance, only its ability to resist Coalition air attack. The Northern Alliance had little in the way of meaningful air assets that posed a threat to the Taliban air defences. Until the concentrated attacks on the Taliban's field forces, the degradation of the Taliban's communications was the greatest hindrance to their conduct of the civil war against the Northern Alliance. The US Special Forces raids on 19 October, against a command and control facility and an airfield near Kandahar also illustrated the Coalition's freedom of operations on the ground. However, it was B-52 carpet-bombing, coupled with Taliban defections and withdrawals, that produced dramatic results and hastened the end of organised resistance. By the end of October, air strikes began to shift away from high-profile urban targets towards Taliban front-line positions.

Heavy strikes, including carpet-bombing B-52s, were conducted on 31 October 2001 against Taliban forces near Bagram, 30 miles north of Kabul. These attacks, lasting several hours, were the most intense against Taliban front-line positions since the air campaign began. The following day, the strategic Taliban garrison at Kala Ata, guarding the approaches to Taloqan, was also attacked. The raid lasted for over four hours and windows were allegedly broken up to 15 miles away. Attacks also continued in the Kandahar and Mazar-e-Sharif areas. Within a week of this intense bombing, the Taliban crumbled first at Mazar-e-Sharif, then Kabul and Jalalabad, their forces were in headlong retreat to Kandahar.

In the face of the Northern Alliance advance, in just a few days during early November the Taliban lost control of much of the country. Their failure was due to a combination of America's air attacks, sizeable Taliban defections and an unprecedented level of co-operation between rival anti-Taliban factions. The Northern Alliance quickly gained control of most of the non-Pashtun cities north of a line extending from Herat, in the west, to Kabul in the east. In total, about 250 Taliban targets suffered air strikes, with the USAF's strategic bombers comprising less than 10 per cent of the force and dropping about 50 per cent of the bombs. By the end of November, with the collapse of the Taliban field forces the focus of the air campaign switched to the Kandahar area and Tora Bora, near Jalalabad in eastern Afghanistan.

Within three months of the air campaign commencing, the Taliban Government had completely collapsed. Osama bin Laden and Al-Qaeda suffered notable losses, particularly after the sieges of Mazar-e-Sharif and Kunduz. The Taliban troops trapped in Kunduz surrendered, leaving some 2,000–5,000 foreign supporters to flee or capitulate. After the fall of Kabul, the Taliban retired to prepared positions in and around Kandahar, their spiritual heartland. American air attacks were then directed against assessed Al-Qaeda terrorist camps in the south of Helmand Province.

Kandahar surrendered to opposition forces on 7 December 2001, without a fight. American air strikes continued against suspected terrorist havens, while British and American Special Forces focused on their search for Osama bin Laden and Taliban leader, Mullah Omar. In the rest of the country, sizeable pockets of Taliban were mopped up.

The Special Forces concentrated on suspected terrorist strongholds, particularly Tora Bora, where Osama was believed to be hiding. US Special Forces did not arrive in the Tora Bora area until 26 November, partly because it had taken the CIA so long to 'acquire' allies on the ground – no doubt using the US dollar. Colonel John Mulholland, commander of Task Force Dagger, was chafing at the bit, but first his forces had to get established. Mohammad Omar and 500 of his supporters were thought to be besieged in the rugged mountains of the Baghran area of northern Helmand. While the Tora Bora region was heavily bombed, Afghan opposition forces under the Eastern Council started to advance into the area. These troops blocked off all escape routes prior to launching a major offensive on the region following the fall of Kandahar.

Remnants of the Taliban's armed forces and Al-Qaeda fighters, perhaps over 1,000 strong, withdrew on the Tora Bora stronghold high in the White Mountains. Their intention seemed to be to conduct hit-and-run attacks or make a last stand if necessary. This defensive complex lay south of the city of Jalalabad and west of the famous Khyber Pass, which links Afghanistan with neighbouring Pakistan. Consisting of miles of interconnecting caves and tunnels, it was built by the Mujahideen who carved the Tora Bora complex into the side of Ghree Khil Mountain with American and Pakistani sponsorship during the Soviet occupation.

Tora Bora first saw action twenty years earlier, in June–July 1981, when a Soviet-led offensive caused widespread destruction of local agriculture, but failed to deliver a decisive result against the Mujahideen. The Soviets heavily bombed the stronghold but could not root out the resistance fighters.

In 2001 the stronghold was believed to be well provisioned with weapons and food that could allow bin Laden and his supporters to withstand a lengthy siege. Russian advice to the Coalition was that Tora Bora could prove to be impregnable if the defenders resisted to the last. The Russians even dusted off their old maps and provided them to the Coalition. Almost the entire American strike force, ten long-range bombers, a dozen tactical jets and four or five gunships concentrated on one complex near Tora Bora over the space of a few days.

The bombing proved successful, with the death of many Al-Qaeda fighters, including twelve senior personnel in early December. A Daisy Cutter, or Big Blue, was dropped on a cave near Tora Bora on 9 December. The weapon was previously deployed in Vietnam and the Gulf War. The 15,000lb BLU-82 was developed in the late 1950s for clearing helicopter landing zones. However, it was hoped the massive pressure wave cause by the blast would kill suspected Al-Qaeda leadership in the cave. The BLU-82 had already been used three times against Taliban and Al-Qaeda front lines, but this was the first time it was used against a hard and deeply buried target.

Pakistan stationed 50,000 troops at the Afghan frontier, of whom 8,000 army and paramilitary troops were sent to stop Taleban or Al-Qaeda fighters crossing the border. For the first time, Pakistan moved the army into the Terah Valley area following reports that Arab suspects might sneak into the tribal territory via Spin Ghar (White Mountain) to flee the intense US bombing on caves in Tora Bora.

Senior military commander in the eastern region, Haji Mohammad Zaman, leading the Eastern Council opposition forces numbering several thousand men, began the laborious task of clearing the Tora Bora valley, before advancing on the stronghold itself. Firstly, they had to capture the village of Tora Bora and other surrounding hamlets, then clear the Tora Bora forest. A bitter battle was expected and for the first week the cautious advance was met with some resistance. The Eastern Council's efforts were sluggish because co-ordination between the various forces was poor, however, Al-Qaeda was slowly driven back higher up into the mountains.

It was anticipated that the Tora Bora stronghold would be protected by minefields, ingenious booby traps and defended by Islamic fanatics prepared to resist to the last. It was potentially a death trap. America stated that it would not deploy its 500 marines on standby at Kandahar, but leave it to its Special Forces and the Eastern Council to clear out the caves. On the Pakistani side of the border, Pakistan's military kept the Khyber and Bati passes closed to try to prevent any terrorists slipping through undetected and then up into the unruly North-West Frontier Province. As many as 2,000 Al-Qaeda fighters were thought to have fled towards Pakistan.

The opposition forces proceeded to pound the mountains with artillery, mortars, rockets and Russian-supplied T-62 tanks. The Al-Qaeda fighters, trapped on all sides by the peaks and valleys of the 15,400ft White Mountains, had lost most of their heavy equipment. They had nothing with which to shoot back at the Eastern Council's exposed tanks perched on the foothills of the White Mountains. In the clear blue skies, American B-52s dropped their bombs in a continuing effort to destroy the caves. To the defenders of Tora Bora, it must have seemed as if the Americans and Eastern Council forces were trying to smash the very mountains themselves.

After two weeks of constant US bombing, the defenders commenced radio negotiations on possible terms. They wanted to surrender to the UN to avoid being handed over to the Americans, but the US would not countenance it. The talks ended and the opposition forces requested the US halt its attacks. They then advanced on the caves still held by Arab Al-Qaeda fighters. The mountain paths were narrow and those caves undetected by the bombers were difficult to locate.

During the second week of December, in the face of varying degrees of resistance, Eastern Council forces captured most of Al-Qaeda's remaining positions. The hunt was now stepped up as American reporter Phil Smucker observed at first hand:

> Just across the ravine were two truckloads of plain clothes Green Berets, who were frantically combing the area. They looked, I judged by their appearance, to be belatedly obsessed with an expired mission. Indeed, Col John Mulholland, commander of Task Force Dagger, which was now engaged in this spelunking exercise, would complain bitterly to me months later about the Promethean mission … he would tell me, 'This was the great cave hunt which I started to refer to as "Operation Vampire" …'

The main ground offensive started late on 13 December. The core of the defenders was thought to number 300, of whom 150 were Arabs, seventy-five Chechens, and the rest Uzbek, Tajik and Afghans. However, other estimates put them at anything from 500 to 1,000 strong.

Bodies of men posted in machine-gun posts along the ridges were found scattered around the scorched remains of their weapons. The first caves captured revealed evidence of desperate attempts to escape the bombing. Strewn with documents, equipment, ammunition, bloodied rags and weapons, their walls were blackened from fire. The Al-Qaeda fighters who had not been killed had fled with no time to plant the feared booby traps.

Phil Smucker wrote:

> The battle for Tora Bora was arguably one of the most exciting battles I ever covered as a correspondent. Above us in the clear, blue sky a white Predator drone circled. It buzzed us for five minutes, hovering with its Hellfire missiles like a bird caught in an updraft, then zipped off to unleash its load … Within thirty minutes we came out of the gorge and saw the first sign of the vast network of caves and bunkers that is known as Tora Bora. These caves had been the redoubt of … Al-Qaeda.

Opposition troops then entered secret hideaways in the Malawa Valley for the first time. The ceilings were made from rock, with walls several feet thick and built from tree trunks packed with stone and dirt. British and US Special Forces came under machine gun and mortar fire as they advanced up the rocky slopes. They then fought fierce battles for control of the area with the survivors. Grouped in two or three areas, the last of the terrorist forces were trapped between two parallel valleys. With opposition forces pushing their way from the north and Pakistani forces blocking the border to the south, the Al-Qaeda fighters could only flee east or west over the mountains or fight their way out.

Three days later, Commander Mohammad Zaman announced that his forces had attacked Spin Ghar (White Mountain) itself at 9 a.m. without any resistance. They came across fifty Al-Qaeda dead and captured one Arab fighter. Zaman claimed to have captured 80 per cent of the mountain. By the evening it was all over, with the capture of thirty-five

Al-Qaeda fighters and the discovery of about 200 dead. The Al-Qaeda suicide squads did not materialise. Despite exhortations to fight to the last, many of the dispirited defenders were buried by the bombardment, surrendered or just slipped away hoping to escape into Pakistan.

In the aftermath of Tora Bora, Pakistan arrested at least 350 Al-Qaeda members, including more than 300 Arab nationals. Most were transferred to the US detention facility, Camp Rhino, at Kandahar. Nonetheless, opposition commanders described the final battle as brutal, leaving the slopes of Tora Bora strewn with the bodies of Al-Qaeda fighters.

American reporter Philip Smucker, with the US forces, recalled:

> Both of the caves I crept into resembled the portals of pharaonic tombs I had entered a year earlier in Egypt's valley of the kings. The booty inside was not as nearly as mesmerizing – only boxes upon boxes of Chinese anti-aircraft shells and Russian rockets, a few spent milk cartons, and a box of Cheerios …
>
> When we arrived back in Upper Pachir, a gaggle of Afghan children surrounded us … They laughed and shouted, 'Osama, Osama Pakistan!' I think I knew what they were getting at, and I pointed across the snow capped Spin Ghar as they nodded and leapt in the air.

By mid-December 2001 the Al-Qaeda supporters had been almost completely driven from the White Mountains. The speed of the victory came as a surprise to everyone. Zaman Gamsharik, opposition forces defence chief in eastern Afghanistan, announced on 16 December 2001, 'This is the last day of Al-Qaeda in Afghanistan. We have done our duty we have cleansed our land of all Al-Qaeda.' US

Defense Secretary Donald Rumsfeld, visiting US forces at Bagram near Kabul, told them, 'The World Trade Center is still burning as we sit here. They are still bringing bodies out. Fortunately, the caves and tunnels at Tora Bora are also burning.' However, despite the victory at Tora Bora, Osama bin Laden had escaped capture.

The USAF launched 138 sorties on 17 December, most of them over Tora Bora and around Kandahar. The following day, it made as many but did not drop any bombs. The battle for Al-Qaeda's Tora Bora stronghold resulted in the loss of some 300 defenders, including Chechens, Pakistanis and Saudis. Two-thirds were killed and the rest taken prisoner. The number of foreigners remaining in the mountains was thought to number between 2,000–5,000 but, scattered, cold and hungry, they were not deemed a threat. By the end of the war the new Afghan Government was holding about 7,000 prisoners.

All that remained after Tora Bora was to mop up. Local opposition forces scoured the caves for clues to the whereabouts of Osama bin Laden. Akram Lahori, of the Lashkar-e-Jhangvi terror group was killed during the fighting in Tora Bora, while his chief, Riaz Basra, was captured in Kandahar. Searchers also found heavy weapons, including one or two tanks, inside some of the caves. US forces had examined seven or eight abandoned cave complexes in the Tora Bora region, where some believed bin Laden had been hiding. By early January 2002 the US had publicly given up chasing him to focus on capturing his remaining supporters.

US air strikes continued against regrouping Taliban/ Al-Qaeda forces and their facilities in eastern Afghanistan. These were concentrated in Paktia Province against the Zhawar Kili cave complex, the site of a Soviet-backed offensive in 1987. The facility first suffered from US air strikes in

1998, in retaliation for the attacks on US embassies in East Africa. According to the US, on 28 December they bombed a walled compound and bunker associated with the Taliban and Al-Qaeda leadership in Paktia Province. The media subsequently claimed the attack killed up to 100 innocent civilians.

Sustained raids were conducted on both Zhawar Kili and anti-aircraft defences near the town of Khost. The US feared that Zhawar Kili was going to be another Tora Bora. General Richard Myers of the US joint chiefs of staff said, 'We have found this complex to be very, very extensive. It covers a large area. When we ask people how large they often describe it as huge.' The camp was composed of three separate training areas and two cave complexes. US Marines and Special Forces moved into the areas after an initial wave of strikes by B-1 and B-52 bombers and carrier-based navy fighters. They then piled up unexploded ammunition and heavy weapons, which were destroyed by a second series of attacks.

Despite three attacks on the complex over a four-day period by US aircraft, surviving Al-Qaeda leaders repeatedly tried to regroup at a warren of caves and bunkers. One strike on the Zhawar Kili training camp hit tanks and artillery, but it was feared the terrorists remained. On 6 January 2002, strikes on Khost and the Zhawar Kili training camp were amongst 118 sorties flown by US air assets over Afghanistan. In one week, 250 bombs were dropped on Zhawar Kili alone.

In response to the attacks on New York's World Trade Center, Operation Freedom had within three months brought the Taliban Government to its knees, ending five years of hated rule and scattering the Al-Qaeda terrorist organisation. Encouraged by the swift resolution in Afghanistan, many countries with longstanding terrorist problems made fresh attempts to address them and there was a greater level of support for the Middle East Peace Process. At the same time, the US developed a broader anti-terrorist strategy focusing on Al-Qaeda links and networks in such countries as the Philippines, Somalia, Uzbekistan and Yemen.

In Afghanistan, the battle at Tora Bora proved not so much a dramatic last stand, but more of a desperate rearguard action to distract the Coalition while the Taliban and Al-Qaeda leadership fled. Al-Qaeda's supporters who were trapped in Tora Bora proved not to be a resilient guerrilla army, as expected, but demoralised terrorists whose plans for global mayhem had gone awry. Despite the success of the air campaign, Afghanistan was not Bosnia, Serbia or Kosovo – the country was not a target-rich environment. For policymakers either side of the Atlantic finding adequate targets that provided strategic leverage against the Taliban proved difficult, if not near impossible. Crucially, without crippling airstrikes against the Taliban's field forces, which aided the opposition Northern Alliance to seize Kabul or Mazar-e-Sharif, the regime was able to shrug off most of the early attacks.

Tora Bora proved a salutary lesson that terrorist havens around the world would not be tolerated and that counter-terrorism and asymmetric warfare had become the dominant type of conflict in the twenty-first century. It was the hunt for Osama bin Laden that gave rise to the drone wars, in which armed, unmanned aerial vehicles were regularly used to eliminate Al-Qaeda's leaders wherever they might be. As for bin Laden, the cause of the battle, he did escape to Pakistan and remained in hiding there until 2011, when US Special Forces finally tracked him down and killed him.

21

Softwar and Cyberwar: The New Battlefields

By the twenty-first century, the warring farmers had come an awfully long way in terms of how they fought their wars. Ironically, of all the enormous technological advances of the late twentieth century, the Vietnam War proved that the media had become one of the most potent weapons. Public opinion had firmly become an instrument of war in its own right. Dramatically, General Ricardo Sanchez, using over-head intelligence imagery of locations Wolverine One and Two south of Tikrit, briefed the world about the capture of Saddam Hussein on 13 December 2003. Amid the cheers, the pictures, almost taken for granted by the press, showed that military imagery had come fully of age in an era of twenty-four-hour global media coverage.

The use of satellite and aerial imagery was no longer confined to the top-secret business of targeting, battle damage assessment and mapping, it was now a valuable asset in the media war – known as information warfare, or 'Softwar'. This is the new battlefield, along with the internet.

After the Second World War the satellite became one of the new weapons, along with nuclear missiles on the front line of the Cold War. The turning point in Washington's perception of the public utility of secret imagery came in 1962 during the Cuban Missile Crisis. Ironically, it was America's obsession with spying on the Soviet Union's nuclear capabilities that had aggravated the situation in the first place.

Their deteriorating relationship, after Gary Powers' U-2 spy plane was shot down over the Soviet Union in 1960, contributed to Moscow's decision to place nuclear missiles on Cuba. Washington detected their presence on 14 October 1962, using unarmed reconnaissance aircraft and confirmed it to the world eight days later. America's UN Ambassador publicly confronted his Soviet counterpart with the evidence and demanded their immediate removal. The world was convinced of Moscow's wrongdoing by the release of these images. Once the crisis was resolved, US aerial reconnaissance of Cuba's Port Mariel on 4 November 1962 confirmed the missiles were being withdrawn. A nuclear war was narrowly averted.

American policymakers were not slow on the uptake, if imagery could shape public opinion over Soviet involvement in Cuba, then it could help justify escalating US involvement in Vietnam. In support of operations, America deployed its tactical reconnaissance aircraft. In the spring of 1965, US intelligence officials received aerial photos of North Vietnamese surface-to-air-missile and anti-aircraft artillery sites. The following year, recon aircraft photographed Soviet MiG-17 interceptors north-east of Hanoi (in mid-1966 North Vietnam only had about sixty-five MiGs, by early 1972 they numbered 200). Photography of the MiG-21 showed that it was appearing in increasing numbers. All this served to prove that Moscow was upping the ante in Vietnam.

Israel was also quick to learn the power of overhead imagery following the Cuban Missile Crisis. Its overwhelming pre-emptive strike at the start of the Six Day War in 1967 successfully pounced on the Egyptian Air Force on the ground. To drive home this triumph, the Israelis issued aerial photography that included destroyed Egyptian fighter jets. Similarly, the Egyptians, after successfully breaching the Israeli sand ramparts along the Suez Canal in 1973, could not help themselves from issuing imagery to the world of their successful bridging operations.

Eighteen years later, during the 1991 Gulf War, British photo interpreters found themselves at odds with their American counterparts. While publicly Washington was issuing hi-tech video of pinpoint bombing raids giving credence to the success of the Coalition's overwhelming air campaign against Saddam, behind the scenes a row developed over its actual effectiveness. This was a case of the information war saying one thing and the analysts another.

General Chuck Horner, US Air Force commander, got exasperated over the issue and at the time was reported saying, 'Unfortunately, the bean-counters of the world absolutely demand accurate data.' Impressively, his fighter-bombers' cameras indicated that they were destroying over 150 Iraqi armoured vehicles and artillery pieces every single night. However, battle damage assessments varied wildly from 10–40 per cent of these figures.

Desert Storm was an imagery war like no other, as the world's population were briefed daily in almost real time about what was happening on the battlefield. Immediately after each raid, edited and specially selected video footage was released showing laser-guided bombs zooming down ventilation shafts amid spectacular pyrotechnics. In theory, American ground units also had direct access to satellite photos. However, one US Army Intelligence officer recalled that he never received a single relevant picture; none showed the ground over which he was advancing. The anonymous officer complained, 'The pictures were supposedly of different areas but I compared them. They were the same place, only one picture was the wrong way up. And this was supposed to be the hi-tech war!' Embarrassingly, some of them had been dealt with by British analysts and were of the same position with different co-ordinates, only the image had been processed upside down.

The global twenty-four-hour news' ability to affect the national will like never before has been dubbed 'Softwar'. This is defined as the hostile use of global media to shape another nation's will by changing its perception of events. In the 1970s, all Hanoi had to do was parade captured USAF pilots or show the results of America's bombing on civilians to the US television networks and there was public outcry (Saddam Hussein repeated this tactic in 1991). After losing the media war in Vietnam to public opinion, Washington was slow in implementing the painful lessons.

In contrast, the Gulf War was the most media-managed conflict in history – that is, until Afghanistan ten years later. The first confirmation of US air attacks on Baghdad in 1991 came from the CNN news channel. General Horner even used their reports to help gauge his airstrikes' effectiveness, despite the fact that CNN could not verify its information. US officers employing CNN to complement their daily updates during Desert Storm did not appreciate that the CNN received in America was different from that broadcast to the rest of the world.

Two years later, General Aidid used CNN to monitor US forces in Somalia. The famously botched mission to capture him, which became known as 'Black Hawk Down', caused a public outcry at home and contributed to the decision to withdraw US forces the following spring. American policy was completely compromised because of the loss of two helicopters and eighteen dead. A former CNN correspondent and US Government media advisor claimed that America's biggest mistake was not releasing the gun video footage showing the relieving helicopter force killing over 1,000 Somalis, this would have made US losses much more palatable.

By the close of the twentieth century, information warfare and Softwar had gone viral. With the rise of the internet during the 1990s, the Americans formulated concepts of cyber warfare. They argued that without robust cyber defences, the state and civil society was at risk as its critical infrastructure could be undermined without a war even breaking out. In the opening decades of the twenty-first century, state-sponsored hacking made cyber warfare a reality. A new arms race was taking place, but not on the traditional battlefield. The generals of the future are faceless hackers who kill software, not people. The conduct of war over the last 1,000 years has come a long, long way since the days of the battleaxe and the shield. The firewall has become the new shield wall.

BIBLIOGRAPHY

Introduction: The Killing Game

Holmes, Richard (ed.), *The World Atlas of Warfare: Military Innovations that Changed the Course of History* (London: Mitchell Beazley, 1988).

Keegan, John, & Richard Holmes, *Soldiers: A History of Men in Battle* (London: Hamish Hamilton, 1985).

Montgomery, Field Marshal, *A Concise History of Warfare* (London: Collins, 1972).

Warry John, *Warfare in the Classical World* (London: Salamander, 1980).

Wise, Terence, *Ancient Armies of the Middle East* (London: Osprey, 1981).

1: The Vikings are Coming: Fulford Gate 1066

Barclay, Brigadier C.N., *Battle 1066* (London: J.M. Dent & Sons, 1966).

Bradbury, Jim, *The Battle of Hastings* (Stroud: Sutton, 1998).

Brøndsted, Johannes, *The Vikings* (Harmondsworth: Penguin, 1978).

Brooke, Christopher, *The Saxon and Norman Kings* (London: B.T. Batsford, 1963/London: Fontana, 1981).

Brooks, F.W., *The Battle of Stamford Bridge* (The East Yorkshire Local History Society, 1956).

Denny, Norman, & Josephine Filmer-Sankey, *The Bayeux Tapestry: The Story of the Norman Conquest 1066* (Glasgow: Collins, 1984).

Furneaux, Rupert, *Conquest 1066* (London: Secker & Warburg, 1966).

Garmonsway, G.N. (trans. & ed.), *The Anglo-Saxon Chronicle* (London: Dent Dutton, 1978).

Glover, R., 'English Warfare in 1066', *The English Historical Review*, Vol. LXVII (1952).

Heath, Ian, *Armies of the Dark Ages 600–1066* (Worthing: Wargames Research Group, 1980).

Heath, Ian, *The Vikings* (London: Osprey, 1985).

Howarth, David, *1066: The Year of Conquest* (Milton Keynes: Robin Clark, 1980).

Laing, Lloyd, & Jennifer, *Anglo-Saxon England* (London: Routledge & Kegan Paul, 1979).

Magnusson, Magnus, & Hermann Palsson (trans.), *King Harald's Saga* (Harmondsworth: Penguin, 1984).

McLynn, Frank, *1066: The Year of the Three Battles* (London: Jonathan Cape, 1998).

Scholfield, G., 'The Third Battle of 1066', *History Today*, Vol. XVI No. 10 (October 1966).

Simpson, Jacqueline, *The Viking World* (London: B.T. Batsford, 1980).

Stenton, Sir Frank, 'Anglo-Saxon England', *The Oxford History of England* (Oxford: Clarendon, 1985).

Thorpe, Lewis, *The Bayeux Tapestry and the Norman Invasion* (London: The Folio Society, 1973).

Walker, Ian W., *Harold: The Last Anglo-Saxon King* (Stroud: Sutton, 1997/Wrens Park, 2000).

Wilson, David, *The Anglo-Saxons* (Penguin: Harmondsworth, 1981).

Wise, Terence, *1066: Year of Destiny* (London: Osprey, 1979).

Wise, Terence, *Saxon, Viking and Norman* (London: Osprey, 1979).

Wood, Michael, *In Search of the Dark Ages* (London: BBC, 1981).

2: Chivalry in Armour: Northallerton 1138

Bradbury, Jim, *Stephen and Matilda: The Civil War of 1139–53* (Stroud: Sutton, 1996).

Davis, R.H.C., *King Stephen 1135–1154* (London: Longman, 1990).

Duggan, Alfred, *Devil's Brood: The Angevin Family* (London: Faber & Faber, 1967).

Forde-Johnston, James, *Great Medieval Castles of Britain* (London: Guild, 1979).

Gascoigne, Christina, *Castles of Britain* (London: Thames & Hudson, 1975).

Pain, Nesta, *Empress Matilda: Uncrowned Queen of England* (London: Weidenfeld & Nicolson, 1978).

Stenton, Doris, May, *English Society in the Early Middle Ages (1066–1307)* (Harmondsworth: Pelican 1951/1955).

Tyerman, Christopher, *Who's Who in Early Medieval England (1066–1272)* (London: Shepheard-Walwyn, 1996).

Warner, Philip, *Sieges of the Middle Ages* (London: Penguin, 2000).

3: Destruction of the Crusaders: Homs 1281

Brent, Peter, *The Mongol Empire – Genghis Khan: His Triumph and his Legacy* (London: Weidenfeld & Nicolson, 1976).

Bridge, Antony, *The Crusades* (London: Granada 1980/Panther 1985).

Gibbon, E., *The History of the Decline and Fall of the Roman Empire, Volume 8* (Folio Society, reprinted 1999).

Heath, Ian, *Armies and Enemies of the Crusades 1096–1291* (Worthing: Wargames Research Group, 1978).

Heath, Ian, *Byzantine Armies 886–1118* (London: Osprey, 1979).

Hyland, Ann, *The Medieval Warhorse: From Byzantium to the Crusades* (Stroud: Sutton, 1994).

Hyland, Ann, *The Warhorse 1250–1600* (Stroud: Sutton, 1998).

Mayer, Hans Eberhard (trans. by John Gillingham) *The Crusades* (Oxford: Oxford University Press, 1978).

Nicolle, David, *Medieval Warfare Source Book: Christian Europe and its Neighbours* (London: Brockhampton, 1998).

Payne, Robert, *The Crusades* (Ware: Wordsworth, 1998).

Robinson, John J., *Dungeon, Fire and Sword: The Knights Templar in the Crusades*. (London: Michael O'Mara, 1994).

Runciman, Steven, *The Crusades, Volume III: The Kingdom of Acre and the Later Crusades* (Cambridge: Cambridge University Press, 1954/ Harmondsworth: Penguin, 1990).

Turnbull, S.R., *The Mongols* (London: Osprey, 1980).

Wise, Terence, *Armies of the Crusades* (London: Osprey, 1978).

Wise, Terence, *The Wars of the Crusades 1096–1291* (London: Osprey, 1978).

4: Triumph of the Bow: Crécy 1346

Brereton, Geoffrey (trans., ed.), *Froissart Chronicles* (Harmondsworth: Penguin, 1978).

Burne, Lt Col Alfred H., *The Crécy War: Military History of the Hundred Years War from 1337 to the Peace of Bretigny 1360* (Ware: Wordsworth, 1999).

Cassell's History of England, Vol. I: From the Roman Invasion to the Wars of the Roses (London: Cassell, 1912).

Lloyd, Alan, *The Hundred Years War* (London: Granada, 1977).

Neillands, Robin, *The Hundred Years War* (London: Guild, 1990).

Nicolle, David, *Italian Medieval Armies 1300–1500* (London: Osprey, 1983).

Nicolle, David, *Medieval Warfare Source Book: Warfare in Western Christendom* (London: Brockhampton, 1995).

Prestwich, Michael, *Armies and Warfare in the Middle Ages: The English Experience* (London: Yale University Press, 1999).

Rothero, Christopher, *The Armies of Crécy and Poitiers* (London: Osprey, 1981).

Seward, Desmond, *The Hundred Years War: The English in France 1337–1453* (London: Constable, 1996).

Tuchman, Barbara W., *A Distant Mirror: The Calamitous 14th Century* (Harmondsworth: Penguin, 1984).

Wailly, Henri de, *Crécy 1346: Anatomy of a Battle* (Poole: Blandford, 1987).

5: When Roses Clash: Bosworth 1485

Bennett, Michael, *The Battle of Bosworth* (Stroud: Alan Sutton, 1993).

Cassell's History of England, Vol. II: From the Wars of the Roses to the Great Rebellion (London: Cassell, 1912)

Cheetham, Anthony, *The Life and Times of Richard III* (London: Weidenfeld & Nicolson, 1992).

Cole, Hubert, *The Wars of the Roses* (London: Hart-Davis, MacGibbon, 1973).

Evans, H.T., *Wales and the Wars of the Roses* (Stroud: Wrens Park, 1998).

Fraser, Antonia (ed.), *The Wars of the Roses* (Berkeley: University of California, 2000).

Gillingham, John, *The Wars of the Roses: Peace and Conflict in Fifteenth-Century England* (London: Weidenfeld & Nicolson, 1993).

Haigh, Philip A., *The Military Campaigns of the Wars of the Roses* (Godalming: Bramley, 1995).

Honigmann, E.A.J. (ed.), William Shakespeare, *King Richard the Third* (London, 1981).

Lander, J.R., *The Wars of the Roses* (Stroud: Sutton, 1992).

Neillands, Robin, *The Wars of the Roses* (London: Brockhampton, 1999).

Rees, D., *The Sun of Prophecy: Henry Tudor's Road to Bosworth* (London: Black Raven, 1985).

Rose, A.L., *Bosworth Field & the Wars of the Roses* (Ware: Wordsworth, 1998).

Weir, Alison, *Lancaster & York: The Wars of the Roses* (London: Pimlico, 1998).

Williams, Neville, *The Life and times of Henry VII* (London: Weidenfeld & Nicolson, 1994).

Wise, Terence, *Medieval European Armies* (London: Osprey, 1975).

Wise, Terence, *The Wars of the Roses* (London: Osprey, 1983).

6: Rape of the New World: Tenochtitlan 1521

Bedini, Silvio A. (ed.), *Christopher Columbus and the Age of Exploration* (New York: Da Capo, 1998).

Burland, C.A., *Peoples of the Sun: The Civilisations of Pre-Columbian America* (New York: Praeger, 1976).

Coe, Michael D., *The Maya* (Harmondsworth: Penguin, 1980).

Coe, Michael D., *Mexico: From the Olmecs to the Aztecs* (London: Thames & Hudson, 1994).

Díaz, Bernal (trans. J.M. Cohen), *The Conquest of New Spain* (Harmondsworth: Penguin, 1963).

Diehl, Richard A., *Tula: The Toltec Capital of Ancient Mexico* (London: Thames & Hudson, 1983).

Elliot, J.H., *Imperial Spain 1469–1719* (Harmondsworth: Penguin, 1978).

Gómara, Francisco López de (trans., ed. Lesley Byrd Simpson), *Cortés: The Life of the Conqueror by his Secretary* (Berkeley: University of California, 1966) (from *Istoria de la Conquista de Mexico*, printed in Zaragoza, 1552).

Hemming, John, *The Conquest of the Incas* (London: Papermac, 1993).

Hordern, Nicholas, *God, Gold and Glory* (London: Reader's Digest, 1980).

Hyatt Verrill, A., *Great Conquerors of South and Central America* (New York: The New Home Library, 1943).

Innes, Hammond. *The Conquistadors* (London: Fontana, 1972).

Meyer, Karl E., *Teotihuacán* (London: Reader's Digest, 1973).

Pendle, George. *A History of Latin America* (London: Penguin, 1990).

Prescott, William H., *History of the Conquest of Mexico* (London: George Routledge & Sons, 1843).

Prescott, William H., *History of the Conquest of Peru* (London: George Routledge & Sons, 1847).

Ross, Kurt, *Codex Mendoza Aztec Manuscript* (Fribourg: Liber, 1978/1984).

Soustelle, Jacques, (trans. Patrick O'Brian), *The Daily Life of the Aztecs* (Harmondsworth: Penguin, 1964).

Stierlin, Henri (ed.), (trans. Marion Shapiro), *Architecture of the World: Ancient Mexico* (Lausanne: Benedikt Taschen, 1967).

Wise, Terence, *The Conquistadores* (London: Osprey, 1980).

Wood, Michael, *Conquistadors* (London: BBC, 2000).

Vaillant, George C., *The Aztecs of Mexico: Origins, Rise and Fall of the Aztec Nation* (Harmondsworth: Penguin, 1951).

7: Bastion of Christendom: St Elmo 1565

Bradford, Ernle, *The Great Siege* (London: Hodder & Stoughton, 1961).

Bradford, Ernle, *The Shield and the Sword: The Knights of Malta* (London: Fontana, 1974).

Correggio, Francesco Balbi di (trans. Balbi, H.A.), *The Siege of Malta* (Copenhagen, 1961).

Duffy, Christopher, *Siege Warfare: The Fortress in the Early Modern World 1494–1660* (London: Routledge & Kegan Paul, 1979).

King, Colonel Sir Edwin, & Sir Harry Luke, *The Knights of St John in the British Realm* (London: St John's Gate, 1967).

Lochhead, Ian C., & T.F.R. Barling, *The Siege of Malta 1565* (London: Literary Services/Melita, 1970).

Olivier, Bridget Cassar Borg., *The Shield of Europe: The Life and Times of La Valette* (Valetta: Progress Press, 1977).

Petrie, Sir Charles, *Philip II of Spain* (London: Eyre & Spottiswoode, 1963).

Robertson, W., *History of the Reign of Charles the Fifth* (London: 1856).

8: No Surrender, No Quarter: Magdeburg 1631

Dupuy, Col Trevor Nevitt, *The Military Life of Gustavus Adolphus: Father of Modern War* (New York: Franklin Watts, 1969).

Gardiner, S.R., *The Thirty Years War 1618–48* (London 1919/New York: Greenwood, 1969).

Langer, Herbert, (trans. C.S.V. Salt), *The Thirty Years War* (Poole: Blandford Press, 1980).

Mann, Golo (trans. Charles Kessler), *Wallenstein: His Life Narrated* (London: André Deutsche, 1976).

Parker, Geoffrey, *Europe in Crisis 1598–1648* (London: Fontana, 1979).

Polisensky, J.V., (trans. Robert Evans), *The Thirty Years War* (London: Nel Mentor, 1971).

Steinberg, S.H., *The Thirty Years War and the Conflict for European Hegemony 1600–1660* (London: Edward Arnold, 1975).

Wagner, Eduard, *European Weapons and Warfare 1618–1648* (London: Octopus, 1979).

Ward, Sir A.W., Prothero, Sir G.W., & Sir Stanley Leathers (eds), *The Cambridge Modern History, Volume IV: The Thirty Years War* (Cambridge: Cambridge University Press, 1934).

Wedgwood, C.V., *The Thirty Years War* (London: Jonathan Cape, 1971).

9: Turning Point: Lostwithiel 1644

Burne, Lt Col Alfred H., & Lt Col Peter Young, *The Great Civil War: A Military History of the First Civil War 1642–1646* (London: Eyre & Spottiswoode, 1959).

Dunn Macray, W. (ed.), *The History of the Rebellion and Civil Wars in England By Edward, Earl of Clarendon*, Vol. III, Books VII–VIII (Oxford: Oxford University Press, 1969).

Edgar, F.T.R., *Sir Ralph Hopton: The King's Man in the West 1642–1652* (Oxford: Oxford University Press, 1968).

Lockyer, Roger (ed.), *Clarendon's History of the Great Rebellion* (Oxford: Oxford University Press, 1967).

Miller, Amos C., *Sir Richard Grenville of the Civil War* (London: Phillimore, 1979).

Ridsdill Smith, Geoffrey, & Dr Margaret Toynbee, *Leaders of the Civil Wars 1642–1648* (Kineton: The Roundwood Press, 1977).

Rogers, Colonel H.C.B., *Battles & Generals of the Civil Wars 1642–1651* (London: Seeley Service, 1968).

Underdown, David, *Somerset in the Civil War and Interregnum* (Newton Abbot: David & Charles, 1973).

Wedgwood, C.V., *The King's War 1641–1647* (London: Fontana, 1977).

Wingfield-Stratford, Esmé, *King Charles the Martyr 1643–1649* (London: Hollis & Carter, 1950).

Woolrych, Austin, *Battles of the English Civil War* (London: Batsford, 1961).

Young, P., & R. Holmes, *The English Civil War, A Military History of the Three Civil Wars 1642–1651* (London: 1974).

Young, Brigadier Peter, *Marston Moor 1644: The Campaign & the Battle* (Kineton: The Roundwood Press, 1970).

Young, Brigadier Peter, & Dr Margaret Toynbee, *Cropredy Bridge 1644: The Campaign and the Battle* (Kineton: Roundwood Press, 1970).

10: Sword v. Musket: Falkirk 1746

Barthorp, Michael, *The Jacobite Rebellions 1689–1745* (London: Osprey, 1982).

Black, Jeremy, *Culloden and the '45* (Stroud: Sutton, 1997).

Chandler, David, *The Art of Warfare in the Age of Marlborough* (London: Batsford, 1976).

Craig, Maggie, *Damn' Rebel Bitches: The Women of the '45* (Edinburgh: Mainstream, 1997).

Erickson, Carolly, *Bonnie Prince Charlie* (London: Robson, 1989).

Hufton, Olwen, *Europe: Privilege and Protest 1730–1789* (Glasgow: Fontana, 1980).

Kemp, Hilary, *The Jacobite Rebellion* (London: Almark, 1975).

Prebble, John, *Culloden* (London: Secker & Warburg, 1961).

Prebble, John, *Mutiny: Highland Regiments in Revolt 1743–1804* (Harmondsworth: Penguin, 1977).

Tomasson, Katherine, & Francis Buist, *Battles of the '45* (London: Batsford, 1962).

11: An Army Divided: Wavre 1815

Becke, A.F., *Napoleon and Waterloo: The Emperor's Campaign with the Armée Du Nord 1815* (London: Greenhill, 1995).

Black, Jeremy, *Waterloo: The Battle that Brought Down Napoleon* (London: Icon, 2011).

Chandler, David, *Dictionary of the Napoleonic Wars* (Ware: Wordsworth, 1999).

Chandler, David (ed.), *Napoleon's Marshals* (London: Cassell, 2000).

Chandler, David, *Waterloo the Hundred Days* (Oxford: Osprey, 1998).

Glover, Michael, *The Napoleonic Wars: An Illustrated History 1792–1815* (London: Batsford, 1979).

Glover, Michael, *Warfare in the Age of Bonaparte* (London: Cassell, 1980).

Hamilton-Williams, David, *The Fall of Napoleon: The Final Betrayal* (London: Brockhampton, 1999).

Hamilton-Williams, David, *Waterloo New Perspectives: The Great Battle Reappraised* (London: Brockhampton, 1999).

Haythornthwaite, Philip J., *Uniforms of Waterloo in Colour 16–18 June 1815* (London: Blandford, 1974).

Hibbert, Christopher, *Waterloo* (Ware: Wordsworth, 1998).

Horne, Alistair, *How Far from Austerlitz? Napoleon 1805–1815* (London: Papermac, 1997).

Howarth, David, *Waterloo: A Near Run Thing* (Moreton-in-Marsh: Windrush, 1999).

Howarth, David, *Waterloo: The Official Guide of the Waterloo Committee* (London: Pitkin Pictorials, 1980).

Longford, Elizabeth, *Wellington: The Years of the Sword* (London: Weidenfeld & Nicolson, 1969).

Maxwell, Sir Herbert, *The Life of Wellington: The Restoration of the Martial Power of Great Britain, Vol II* (London: Sampson Low, 1900).

McLynn, Frank, *Napoleon* (London: Jonathan Cape, 1997).

Neillands, Robin, *Wellington and Napoleon: Clash of Arms 1807–1815* (London: John Murray, 1994).

Pericoli, Ugo, & Michael Glover, *1815: The Armies at Waterloo* (London: Seeley Service, 1979).

Rogers, Colonel H.C.B., *Napoleon's Army* (London: Purnell, 1974).

Weller, Jac, *Wellington at Waterloo* (London: Greenhill, 1998).

12: Spear v. Rifle: Isandhlwana 1879

Barthorp, Michael, *The Zulu War: A Pictorial History* (Poole: Blandford, 1980).

Becker, Peter, *Rule of Fear: The Life and Times of Dingane, King of the Zulu* (London: Longmans, 1964).

Binns, C.T., *The Warrior People: Zulu Origins, Customs and Witchcraft* (London: Robert Hale, 1975).

Clammer, David, *The Zulu War* (Newton Abbot: David & Charles, 1973).

Emery, Frank, *The Red Soldier: Letters from the Zulu War 1879* (London: Hodder & Stoughton, 1977).

Featherstone, Donald, *Weapons & Equipment of the Victorian Soldier* (Poole: Blandford Press, 1978).

Jackson, F.W.D., 'Isandhlwana, 1879 – The Sources Re-examined', *Journal of the Society for Army Historical Research*, Nos 173, 175 & 176 (1965).

Knight, Ian, *The Anatomy of the Zulu Army from Shaka to Cetshwayo 1818–1879* (London: Wrens Park, 1999).

Knight, Ian, *Zulu Rising: The Epic Story of Isandlwana and Rorke's Drift* (London: Pan, 2011).

Laband, John, *The Rise & Fall of the Zulu Nation* (London: Arms and Armour Press, 1997).

Lloyd, Alan, *The Zulu War 1879* (London: Hart-Davis MacGibbon, 1973/Newton Abbot: Reader's Union, 1976).

McBride, Angus, *The Zulu War* (London: Osprey, 1976).

Morris, Donald R., *The Washing of the Spears* (London: Jonathan Cape, 1972).

Ritter, E.A., *Shaka Zulu: The Rise of the Zulu Empire* (London: Longman, 1955).

Selby, John, *Shaka's Heirs* (London: George Allen & Unwin, 1971).

Wilkinson-Latham, Christopher, *Uniforms & Weapons of the Zulu War* (London: B.T. Batsford, 1978).

13: Tank v. Tank: Villers-Bretonneux 1918

Forty, George, *Tank Action: From the Great War to the Gulf* (Stroud: Sutton, 1995).

Holmes, Richard, *The Western Front* (London: BBC Worldwide, 1996).

Liddell Hart, B.H., *History of the First World War* (London: Cassell, 1982).

Lloyd George, David, *War Memoirs, Vol. II* (London: Odhams Press, 1937).

Overy, Richard (ed. consultant), *World War I Definitive Visual Guide: From Sarajevo to Versailles* (London: Dorling Kindersley, 2014).

Philpott, William, *Attrition: Fighting the First World War* (London: Abacus, 2015).

Stevens, Philip, *The Great War Explained* (Barnsley: Pen & Sword Military, 2014).

Strachan, Huw, *The First World War* (London: Simon & Schuster, 2003).

Terraine, John, *The First World War 1914–18* (London: Leo Cooper, Secker & Warburg, 1983).

14: Blitz but no Krieg – Britain 1940

Butler, J.R.M., *Grand Strategy, September 1939–June 1941*, Volume II (London: 1957).

Collier, Richard, *Eagle Day: The Battle of Britain* (London: J.M. Dent & Sons, 1980).

Deighton, Len, *Blitzkrieg: From the Rise of Hitler to the Fall of Dunkirk* (London: Jonathan Cape, 1979).

Deighton, Len, *Fighter: The True Story of the Battle of Britain* (London: Jonathan Cape, 1977).

Fleming, Peter, *Invasion 1940* (London: Rupert Hart-Davis, 1957).

Franks, Norman, *The Battle of Britain* (London: Bison, 1981).

Harris, Sir Arthur, *Bomber Offensive* (London: Collins, 1947).

Liddell Hart, B.H., *The Other Side of the Hill* (London: Pan, 1983).

Parkinson, Roger, *Dawn on our Darkness: The Summer of 1940* (London: Granada, 1977).

Trevor-Roper, H.R. (ed.), *Hitler's War Directives 1939–45* (London: Pan, 1983).

Warner, Philip, *Invasion Road* (London: Cassell, 1980).

Wheatly, Ronald, *Operation Sea Lion: German Plans for the Invasion of England 1939–42* (Oxford: Oxford University Press, 1962).

15: Storming the Beaches: D-Day 1944

Belchem, Major General David, *Victory in Normandy* (London: Chatto & Windus, 1981).

Belfield, Eversley, & H. Essame, *The Battle for Normandy* (London: Pan, 1983).

Bruce, George, *Second Front Now! The Road to D-Day* (London: Macdonald & Jane's, 1979).

Carell, Paul, *Invasion: They're Coming!* (London: Corgi, 1963).

D'Este, Carlo, *Decision in Normandy* (London: Pan, 1984).

Hastings, Max, *Overlord: D-Day and the Battle for Normandy* (London: Michael Joseph, 1984).

Hunt, Robert, & David Mason, *The Normandy Campaign* (London: Leo Cooper, 1976).

Johnson, Garry, & Christopher Dunphie, *Brightly Shone the Dawn: Some Experiences of the Normandy Invasion* (London: Frederick Warne, 1980).

Keegan, John, *Six Armies in Normandy* (Harmondsworth: Penguin, 1983).

Lefèvre, Eric, *Panzers in Normandy: Then and Now* (London: After the Battle, 1983).

Lucas, James, & James Barker, *The Killing Ground* (London: B.T. Batsford, 1978).

Maule, Henry, *Caen: The Brutal Battle and the Break-out from Normandy* (Newton Abbot: David & Charles, 1976).

McKee, Alexander, *Caen: Anvil of Victory* (London: Souvenir Press, 1964).

Ryan, Cornelius, *The Longest Day: June 6 1944* (London: Victor Gollanz, 1982).

Seaton, A., *The Fall of Fortress Europe 1943–1945* (London: 1981).

Thompson, R.W., *D-Day: Spearhead of Invasion* (London: Macdonald & Co., 1968).

Turner, John Frayn, *Invasion '44* (Shrewsbury: Airlife, 2002).

Tute, Warren, Costello, John, & Terry Hughes, *D-Day* (London: Pan, 1975).

Warner, Philip, *The D-Day Landings* (London: William Kimber, 1980).

Wilmot, H.P., *June 1944* (Poole: Blandford, 1984).

Wynn, Humphrey, & Susan Young, *Prelude to Overlord* (Shrewsbury: Airlife, 1983).

Young, Brigadier Peter, *D-Day* (London: Bison, 1981).

16: The Great River Crossing: The Rhine 1945

Chalfont, Alun, *Montgomery of Alamein* (London: Weidenfeld & Nicolson, 1976).

Guingand, Sir Francis de, *Operation Victory* (London: Hodder & Stoughton, 1947).

Hamilton, Nigel, *Monty: The Field Marshal 1944–1976* (London: Hamish Hamilton, 1986).

Hastings, Max, *Armageddon: The Battle for Germany 1944–45* (London: Macmillan, 2004).

Horne, Alastair, *The Lonely Leader: Monty 1944–1945* (London: Pan, 1995).

Horrocks, Lt General Sir Brian, *A Full Life* (London: Collins, 1960).

Horrocks, Sir Brian, Belfield, Eversley, & Major General H. Essame, *Corps Commander* (London: Methuen, 1979).

Montgomery, Bernard Law, *The Memoirs of Field Marshal the Viscount Montgomery of Alamein* (London: Collins, 1958).

Thompson, Julian, *The Imperial War Museum Book of Victory in Europe* (London: Sidgwick & Jackson, 1994).

Whiting, Charles, *Bounce the Rhine* (London: Leo Cooper, 1985).

Wilmot, Chester, *The Struggle for Europe* (London: Collins, 1952).

17: A Very Modern Siege: Khe Sanh 1968

Bonds, Ray (ed.), *The Vietnam War: The Illustrated History of the Conflict in Southeast Asia* (London: Salamander, 1981).

Cawthorne, Nigel, *Vietnam: A War Lost and Won* (London: Arcturus, 2003).

Dunstan, Simon, *Armour of the Vietnam Wars* (London: Osprey, 1985).

Karnow, Stanley, *Vietnam, A History: The First Complete Account of Vietnam at War* (Harmondsworth: Penguin, 1984).

Katcher, Philip, *Armies of the Vietnam War 1962–75* (London: Osprey, 1980).

O'Balance, Edgar, *The Wars in Vietnam 1954–73* (Shepperton: Ian Allan, 1975).

Robinson, Anthony (ed.), *Weapons of the Vietnam War* (London: Bison, 1983).

Russell, Lee E., *Armies of the Vietnam War (2)* (London: Osprey, 1983).

Starry, General Donn A., *Armoured Combat in Vietnam* (Poole: Blandford, 1981).

Thompson, Leroy, *Uniforms of the Indo-China and Vietnam Wars* (Poole: Blandford, 1984).

Welsh, Douglas, *The History of the Vietnam War* (London: Bison, 1981).

Wintle, Justin, *The Vietnam Wars* (London: Weidenfeld & Nicolson, 1991).

Woodruff, M.W., *Unheralded Victory: Who won the Vietnam War?* (Arlington 1999/London 2000).

18: Triumph in the Air: Bekaa Valley 1982

Katz, Samuel M., *Armies in Lebanon 1982–84* (London: Osprey, 1985).

Laffin, John, *The War of Desperation: Lebanon 1982–85* (London: Osprey, 1985).

Lambeth, B.S., *Moscow's Lessons from the 1982 Lebanon Air War* (Santa Monica: Rand, 1984).

McDowall, D., *Lebanon: A Conflict of Minorities* (London: 1986).

Pivka, Otto von, *Armies of the Middle East* (Cambridge: Patrick Stephens, 1979).

Schiff, Ze'ev, *A History of the Israeli Army: 1874 to the Present Day* (London: Sidgwick & Jackson, 1987).

Schiff, Ze'ev & Ehud Ya'ari, *Israel's Lebanon War* (London: Unwin, 1986).

Tucker, A., 'The Strategic Implications of the Iran–Iraq War', *Middle East Strategic Studies Quarterly*, Vol. 1, No. 2 (Spring/Summer 1989).

Wright, Robin, *Sacred Rage: The Wrath of Militant Islam* (London: André Deutsch, 1986).

19: A Show of Force: Khafji 1991

Billière, General Sir Peter de la, *Storm Command: A Personal Account of the Gulf War* (London: Harper Collins, 1992).

Brown, Ben, & David Shukman, *All Necessary Means: Inside the Gulf War* (London: BBC Books, 1991).

Bulloch, John, & Harvey Morris, *Saddam's War: The Origins of the Kuwait Conflict and the International Response* (London: Faber & Faber, 1991).

Cordingley, Major General Patrick, *In the Eye of the Storm: Commanding the Desert Rats in the Gulf War* (London: Hodder & Stoughton, 1996).

Dewar, Colonel Michael, *War in the Streets: The Story of Urban Combat from Calais to Khafji* (Newton Abbot: David & Charles, 1992).

Finlan, Alastair, *The Gulf War 1991* (Oxford: Osprey, 2003).

Freedman, Lawrence, & Efrain Karsh, *The Gulf Conflict 1990–1991* (London: Faber & Faber, 1993).

Rottman, Gordon, *Armies of the Gulf War* (London: Osprey, 1993).

Schwarzkopf, General H. Norman, & Peter Petre, *The Autobiography: It Doesn't take a Hero* (London: Bantam, 1993).

Sultan, HRH General Khaled bin, & P. Searle, *Desert Warrior: A Personal View of the Gulf War by the Joint Forces Commander* (London: Harper Collins, 1995).

Watson, Bruce W. (ed.), *Military Lessons of the Gulf War* (London: Greenhill, 1991).

20: Asymmetric Warfare: Tora Bora 2001

Baxter, Jenny, & Malcolm Downing, *The Day that Shook the World: Understanding September 11* (London: BBC Worldwide, 2001).

Bergen, Peter L., *Holy War Inc: Inside the Secret World of Osama bin Laden* (London: Phoenix, 2002).

Bodansky, Yossef, *Bin Laden: The Man who Declared War on the World* (New York: Prima, 2001).

Burke, Jason, *Al-Qaeda: The True Story of Radical Islam* (London: Penguin, 2004).

Pettiford, Lloyd, & David Harding, *Terrorism: The New World Order* (London: Arcturus, 2003).

Rashid, Ahmed, *Jihad: The Rise of Militant Islam in Central Asia* (New Haven & London: Yale Nota Bene, 2003).

Rashid, Ahmed, *Taliban: The Story of the Afghan Warlords* (London: Pan, 2001).

Reeve, Simon, *The New Jackals: Ramzi Yousef, Osama bin Laden and the Future of Terrorism* (London: André Deutsch, 1999).

Robson, Adam, *Bin Laden: Behind the Mask of a Terrorist* (Edinburgh: Mainstream, 2000).

Smucker, Philip, *Al-Qaeda's Great Escape: The Military and the Media on Terror's Trail* (Washington DC: Potomac, 2004).

Tucker-Jones, Anthony, *The Rise of Militant Islam: An Insider's View of the Failure to Curb Global Jihad* (Barnsley: Pen & Sword Military, 2010).

21: Softwar and Cyberwar: The New Battlefields

Campen, Alan D., Dearth, Douglas H., & Thomas R. Gooden
(eds), *Cyberwar: Security, Strategy and Conflict in the Information Age*
(Fairfax, Virginia: Armed Forces Communications and Electronics
Association, 1996).

Campen, Alan D., & Douglas H. Dearth (eds), *Cyberwar 2.0: Myth,
Mysteries and Reality* (Fairfax: AFCEA International, 1998).

Hastings, Max, *Editor: An Inside Story of Newspapers* (London:
Pan, 2003).

Junger, Sebastian, *War* (London: Fourth Estate, 2010).

Potts, David (ed.), *The Big Issue: Command and Combat in the
Information Age* (Command and Control Research Program, 2003).

ACKNOWLEDGEMENTS

It all started because I grew up not far from the Civil War battlefield at Cheriton. I regularly made pilgrimages to the rolling fields that had hosted the clash of arms so long ago. In later life this was followed by visits to Hastings, Landsdown, Roundway Down and many others. At Lostwithiel the remains of Restormel Castle are an enduring monument to another age and a king who lost his crown. I have stood atop the ramparts overlooking Grand Harbour and marvelled at the thought of the Sultan's Janissaries toiling in the heat toward the guns of St Elmo. In the Palace of the Grand Masters, Matteo Perez D'Aleccio's remarkable frescos still vividly bring the siege of Malta alive.

From Lion Hill in the midst of Wellington's line at Waterloo I looked south to Napoleon's positions and east toward Papelotte and Wavre. So much rested on the Prussians that day. From the cliffs of Dover I wondered why, following Dunkirk, Hitler did not cross that narrow strip of water to attack England when it was at its lowest and weakest. Behind Omaha Beach I walked amongst the well-kept but rather melancholy American graves and realised what a triumph D-Day had been.

A large number of people helped with the preparation of this book and thanks are accorded to them all – in particular Chrissy McMorris, commissioning editor at The History Press, who was very supportive of its broad canvass. It is not every editor who is bold enough to indulge an author covering 1,000 years of warfare in just twenty campaigns. Also at THP, I am grateful to my project editor Jezz Palmer; copy editors Sarah Wright and Andrew Bate; and designer Glad Stockdale, who produced the numerous maps. Fellow writers and authors Andy Evans and Tim Newark were instrumental in encouraging some of the initial research for a number of the battles. Likewise, thanks are due to Duncan Macfarlane and Iain Dickie for having encouraged a very early interest in history's great battles, which became a lifelong passion.

INDEX

The destination for history
www.thehistorypress.co.uk